Studies in Computational Intelligence

Volume 913

The series "Studies in Computational Intelligence" (SCI) publishes new developments and advances in the various areas of computational intelligence—quickly and with a high quality. The intent is to cover the theory, applications, and design methods of computational intelligence, as embedded in the fields of engineering, computer science, physics and life sciences, as well as the methodologies behind them. The series contains monographs, lecture notes and edited volumes in computational intelligence spanning the areas of neural networks, connectionist systems, genetic algorithms, evolutionary computation, artificial intelligence, cellular automata, self-organizing systems, soft computing, fuzzy systems, and hybrid intelligent systems. Of particular value to both the contributors and the readership are the short publication timeframe and the world-wide distribution, which enable both wide and rapid dissemination of research output.

The books of this series are submitted to indexing to Web of Science, EI-Compendex, DBLP, SCOPUS, Google Scholar and Springerlink.

More information about this series at http://www.springer.com/series/7092

Aparajita Nanda · Nisha Chaurasia
Editors

High Performance Vision Intelligence

Recent Advances

Springer

Editors
Aparajita Nanda
Jaypee Institute of Information Technology
Noida, India

Nisha Chaurasia
Jaypee Institute of Information Technology
Noida, India

ISSN 1860-949X ISSN 1860-9503 (electronic)
Studies in Computational Intelligence
ISBN 978-981-15-6843-5 ISBN 978-981-15-6844-2 (eBook)
https://doi.org/10.1007/978-981-15-6844-2

This Springer imprint is published by the registered company Springer Nature Singapore Pte Ltd.
The registered company address is: 152 Beach Road, #21-01/04 Gateway East, Singapore 189721, Singapore

Books may look like nothing more than words on a page, but they are actually an infinitely complex imaginotransference technology that translates odd, inky squiggles into pictures inside your head.

—Jasper Fforde

This is the whole point of technology. It creates an appetite for immortality on the one hand. It threatens universal extinction on the other. Technology is lust removed from nature.

—Don DeLillo

Preface

This is the first book of its kind dedicated to the challenges in vision intelligence, including high performance computing applications. The text provides an in-depth, multidisciplinary discussion of recent advancements and state-of-the-art methods in spectrum of disciplines. The Topics include the best research fields ranging from image processing, computer vision, artificial intelligence, machine learning to cloud computing, Internet of Things (IoT), and Big Data Analytics. The content of this book is categorized into Generation and Display, Processing & Analysis, Implementations & Architectures, and Applications.

This book encourages aspirants to share their beliefs and experiences in solving high performance computing problems. The chapters highlight in-depth discussions on emerging cross-disciplinary themes and its applications in vision, intelligence, and high performance computing. This book provides a platform to figure out the significant impact in science, society, research, and industry. It not only helps to connect different research communities in a forum to represent state of the art in domain-specific platforms and technologies, but also supports vision to the advancement of sustainable development focusing smart societies.

This book covers a diverse set of research in cross-disciplinary themes and its publication is ideally designed for academicians, technology professionals, students, and researchers interested in uncovering the latest innovations. It depicts all aspects of image analysis from the low level (early vision aspects) to the high level (recognition and interpretation aspects). Demonstration of ground-breaking issues and paradigms, frameworks, and implications are depicted through charts and graphs. It also features explanatory, illustration of Algorithms, architecture, applications, software systems, and data analytics in the scope of specified domain.

Noida, U.P., India — Dr. Aparajita Nanda
January 2020 — Dr. Nisha Chaurasia

Acknowledgements We, Dr. Aparajita Nanda and Dr. Nisha Chaurasia, wholeheartedly want to thank our family members in helping and guiding us in any manner. Also, we wish to thank each and every individual who helped us in editing this book. Additionally, we are highly grateful to our contributors and reviewers who aided in making of the book.

Contents

Editors and Contributors

About the Editors

Dr. Aparajita Nanda is currently an assistant professor at the Department of Computer Science Engineering, Jaypee Institute of Information Technology, Noida, India. She obtained her Bachelors (IT) from Biju Pattnaik University, Odisha, Masters (CSE) from Sambalpur University Institute of Information Technology, and Ph.D. from the National Institute of Technology Rourkela, India. Her major research interests include computer vision, data mining, machine learning and visual surveillance in computer engineering applications. She has published several papers in respected international journals, and co-authored research articles in conferences, and book chapters. Dr. Nanda is a member of the ACM and IAENG. She has worked on a R&D project funded by SERB.

Dr. Nisha Chaurasia is currently an assistant professor at the Department of Computer Science Engineering, Jaypee Institute of Information Technology, Noida, India. She obtained her Bachelors (IT) and Masters (CSE) from Madhav Institute of Technology and Science, Gwalior, and Ph.D. from ABV-IIITM Gwalior, India. Her research interests include data mining, cloud computing, high-performance computing and its applications. She has published several papers in respected international journals and conference proceedings, and also contributed book chapters. Dr. Chaurasia is a member of the ACM and IAENG. She has worked on a R&D project under UKIERI.

Contributors

Tarushi Agrawal Computer Science & Engineering Department, Jaypee Institute of Information Technology, Noida, India

Gajendra Kumar Ahirwar Department of Information Technology, Bhopal, Madhya Pradesh, India

Vaibhav Aren Department of CSE and IT, Jaypee Institute of Information Technology, Noida, India

Dinesh Bhardwaj Department of Electronics and Communication, Thapar Institute of Engineering and Technology, Patiala, India

Mahua Bhattacharya ABV-Indian Institute of Information Technology and Management, Gwalior, India

Priya Bist Computer Science & Engineering Department, Jaypee Institute of Information Technology, Noida, India

A. Chakrabarti A.K.C.S.I.T., Calcutta University, Kolkata, India

S. Chakraborty A.K.C.S.I.T., Calcutta University, Kolkata, India

Rashmi Chaudhry Department of Computer Science and Engineering, SPM-International Institute of Information Technology Naya Raipur, Naya Raipur, India

N. Chaurasia Department of Computer Science and Engineering, Uttarakhand Technical University, Dehradun, Uttarakhand, India

Meenakshi Das School of Computer Science and Engineering, VIT Vellore, Vellore, India

L. Davila Nicanor University Center UAEM Valley of Mexico, Autonomous University of Mexico State, Boulevard University, Atizapan de Zaragoza, Mexico State, Mexico

P. Dimri Department of Computer Science and Applications, G.B. Pant Engineering College, Pauri, India

R. Ghosh A.K.C.S.I.T., Calcutta University, Kolkata, India

Charu Goyal Department of CSE and IT, Jaypee Institute of Information Technology, Noida, India

K. K. Gupta Department of Information Technology, Rustamji Institute of Technology, Gwalior, India

Madhuri Gupta Computer Engineering and Information Technology, ABES Engineering College (Affiliated with Abdul Kalam Technical University, Lucknow), Ghaziabad, Uttar Pradesh, India

Raahat Gupta ABV-Indian Institute of Information Technology and Management, Gwalior, India

Sarishty Gupta Department of CSE and IT, Jaypee Institute of Information Technology, Noida, India

Kusum Lata Jain CCE, Manipal University Jaipur, Rajasthan, India

Nimit Jain Computer Science & Engineering Department, Jaypee Institute of Information Technology, Noida, India

Anil Kumar Kushwah Department of Computer Science and Engineering, Maulana Azad National Institute of Technology, Bhopal, Madhya Pradesh, India

Varsha Kushwah Department of Computer Science and Engineering, Maulana Azad National Institute of Technology, Bhopal, Madhya Pradesh, India

V. M. Landassuri Moreno University Center UAEM Valley of Mexico, Autonomous University of Mexico State, Boulevard University, Atizapan de Zaragoza, Mexico State, Mexico

Ansuman Mahapatra Department of Computer Science and Engineering, National Institute of Technology, Karaikal, Puducherry, India

Akshat Maheshwari ABV-Indian Institute of Information Technology and Management, Gwalior, India

Saumil Maheshwari ABV-Indian Institute of Information Technology and Management, Gwalior, India

Alekha Kumar Mishra Department of Computer Applications, National Institute of Technology Jamshedpur, Jamshedpur, India

Ambuj Mishra ABV-Indian Institute of Information Technology and Management, Gwalior, India

Smaranika Mohapatra SCE, Poornima University Jaipur, Rajasthan, India

H. R. Orozco Aguirre University Center UAEM Valley of Mexico, Autonomous University of Mexico State, Boulevard University, Atizapan de Zaragoza, Mexico State, Mexico

S. Priyadharshini Department of Computer Science and Engineering, National Institute of Technology, Puducherry, India

Megha Rathi Computer Science & Engineering Department, Jaypee Institute of Information Technology, Noida, India

Pankaj K. Sa Department of Computer Science and Engineering, National Institute of Technology, Rourkela, Odisha, India

Sowmya Saraswathi School of Computer Science and Engineering, VIT Vellore, Vellore, India

Priyansh Saxena ABV-Indian Institute of Information Technology and Management, Gwalior, India

S. H. Shaikh CSE, BML Munjal University, Gurugram, India

Mukta Sharma ABV-Indian Institute of Information Technology and Management, Gwalior, India

Shivani Sharma Department of Computer Science and Engineering, SPM-International Institute of Information Technology Naya Raipur, Naya Raipur, India

Asis Kumar Tripathy School of Information Technology and Engineering, VIT Vellore, Vellore, India

Rahul Verma ABV-Indian Institute of Information Technology and Management, Gwalior, India

Rajesh Wadhvani Department of Computer Science and Engineering, Maulana Azad National Institute of Technology, Bhopal, Madhya Pradesh, India

Acronyms

ABC	Artificial Bee Colony Optimisation
ACA	Average Classification Accuracy
ACO	Ant Colony Optimisation
ADS	Automatic Detection System
AI	Artificial Intelligence
AIML	Artificial Intelligence Markup Language
ALEXA	Artificial Language Experimental Assistant/Amazon LEXicon Assistant
ALICE	Artificial Linguistic Internet Computer Entity
ANN	Artificial Neural Network
AODV	Adaptive On-Demand Vector routing
AR	Autoregressive
ARIMA	Autoregressive Integrated Moving Average
ARIMAX	ARIMA model with exogenous variable X
ARMA	Autoregressive Moving Average
ASR	Automatic Speech Recognition
AUC	Area Under the Curve
AWS	Amazon Web Services
BC	Breast Cancer
BoF	Bag-of-Features
BoW	Bag-of-Words
CAD	Computer Aided Design
CART	Classification And Regression Tree
CC	Cloud Computing
CCP	Cloud Computing Platforms
CD	Cell Decomposition
CDG	Concern Dependence Graph
CE	Context Encoders
CEP	Complex Events Processing
CMP	Cube map Projection

CNN	Convolutional Neural Network
cSWAP	Controlled SWAP
CT	Completion Time
DAG	Directed Acyclic Graph
DASH	Dynamic Adaptive-Streaming over HTTP
DoF	Degree of Freedom
DoG	Difference of Gaussians
DVFS	Dynamic Voltage and Frequency Scaling
DYMO	Dynamic Mobile ad-hoc network
EBM	Evidence-Based Medicine
ELM	Extreme Learning Machine
ERP	Equirectangular Projection
ET	Execution Time
FA	Firefly Algorithm
FFT	Fast Fourier Transform
FoV	Field-of-view
FRQI	Flexible Representation of Quantum Images
GA	Genetic Algorithm
GAN	Generative Adversarial Network
GAS	Generalized Autoregressive Score
GASA	Genetic Algorithm Simulated Annealing
GASX	GAS model with exogenous variable X
GSO	Glowworm Swarm Optimisation
HDR	High Dynamic Range
HMD	Head Mount Device
HOG	Histogram of Oriented Gradient
IaaS	Infrastructure as a Service
ICIAR	International Conference on Image Analysis and Recognition
ID3	Iterative Dichotomiser 3
IoT	Internet of Things
IR	Infrared
IT	Information Technology
JCAHO	Joint Commission on Accreditation of Healthcare Organizations
LACO	Honeybee Cost-Based Efficiency Comparison
LCL	Lower Control Limit
LSVRC	Large Scale Visual Recognition Challenge
MA	Moving Average
MAE	Mean Absolute Error
MANET	Mobile Ad-hoc Network
MCN	Multi-Camera Network
MCQI	Multichannel representation of Quantum Images
ML	Machine Learning
MLP	Multi-Layer Perceptron
NEQR	Novel Enhanced Quantum Representation of Digital Image Model
NIA	Nature-Inspired Algorithms

NLP	Natural Language Processing
NREL	National Renewable Energy Laboratory
PaaS	Platform as a Service
PCA	Principal Component Analysis
PCA-ID	Principle Component Analysis Image Denoising
PFM	Potential Field Method
PI	Progressive inpainting
PoV	Points of View
PRM	Probabilistic Road-Map
PSNR	Peak Signal-to-Noise Ratio
PSO	Particle Swarm Optimization
QER	Quality Emphasized Region
QFT	Quantum Fourier Transform
QoE	Quality of Evaluation
QoS	Quality of Service
QWT	Quantum Wavelet Transform
QWT^{-1}	Inverse Quantum Wavelet Transform
ReLu	Rectified Linear activation function
ResNet	Residual Network
RMSE	Root Mean Squared Error
RNN	Recurrent Neural Networks
RRT	Rapidly-exploring Random Trees
RT	Response Time
RTPSO-B	Ranging function and tuning function-based PSO
SCTMM	Sphere-shaped Coordinates Transform placed Mobility Method
SG	Sub-Goal
SI	Swarm Intelligence
SIFT	Scale Invariant Feature Transform
SIoT	Social Internet of Things
SIRI	Speech Interpretation and Recognition Interface
SP	Stream Processing
SRE	Software Reliability Engineering
SSA	Salp Swarm Algorithm
SSD	Solid State Drive
SSIM	Structural Similarity Index
SURF	Speeded Up Robust Features
SVM	Support Vector Machine
SWRL	Semantic Web Rule Language
TS	Tabu Search
TSS	Trust Score of Sensors
TSV	Trust Score of sensed Value
UAV	Unmanned Aerial Vehicle
UCL	Upper Control Limit

UIMA	Unstructured Information Management Architecture
VAR	Vector Autoregressive
VGGNet	Visual Geometry Group Network
VM	Virtual Machine
WMN	Wireless Mess ad-hoc Network
WSN	Wireless Sensor Networks
WWW	World Wide Web

Social Internet of Things: Opportunities and Challenges in Developing Countries

Kusum Lata Jain and Smaranika Mohapatra

Abstract Internet of Things (IoT) is a technological innovation which probable as the future of information technology for solution and optimization of many real-world problems with number of connected device through a communication media to share monitoring information of environment. A wide spectrum of independent research activities are going on to discover the opportunities of integrating social networking concepts of sharing information to other, into Internet of Things (IoT). The output of this integration, named Social Internet of Things (SIoT) provides a simple communication methods among humans and machines which majorly affects the live hood and culture of living of humans. A continues development in new technologies the method of interaction of human and machine are changed from text to voice, from touch to gesture and from pen to bio-matrix. This can be enhanced by exploiting the social behavior of the users of such devices and offers prospects of new ways of monitoring status, learning, and responding in real time. This paper presents an outline of the enablement of SIoT and their existence in developing countries, general applications, opportunities, and challenges as well as future use of SIoT in these countries is explored.

1 Introduction

We are the witness to the rapid growth in adoption of Internet of Things (IoT) and Social media technologies. Traditionally Internet of Things (IoT) paradigm evolved which connected the various things to Internet and enabled the creation of applica-

K. L. Jain (✉)
CCE, Manipal University Jaipur, Rajasthan 303007, India
e-mail: kusumlata.jain@jaipur.manipal.edu

S. Mohapatra
SCE, Poornima University Jaipur, Rajasthan 303905, India
e-mail: smaranika.mohapatra@poornima.edu.in

A. Nanda and N. Chaurasia (eds.), *High Performance Vision Intelligence*,
Studies in Computational Intelligence 913,
https://doi.org/10.1007/978-981-15-6844-2_1

Fig. 1 Relation of things, people, and cloud service for collective system

tions that helps to solve routine monitoring activities of surroundings more effectively and efficiently [1]. IoT is defined as a system consisting of mechanical and digital machine called smart objects which interact with environment and other nearby objects and possess the information processing abilities, collect information from the same, process it to typically activate one or more digital control signals, e.g., the automatic reordering of groceries or other things, when stock goes below a certain level, to make life easier and convenient. To sense and collect the information from the surrounding, a physical object is the basic essential of IoT, and this is abstracted as a service.

As our phones contain different sensors that constantly record and transmit enormous amounts of information without getting noticed or being aware of it; homes, cars are "smarter" than ever before; our public infrastructure such as street lights, elevators, escalators, roads, etc. contains innumerable sensors that are essential for their maintenance and our safety; and factories even those producing low-tech products have begun to embrace the "Industrial Internet" which is powered by IoT. IoT has established a world where humans are provided with smart data services by the fusion of physical objects and information networks [2]. This paradigm is leading toward to provide better living conditions for mankind, where the smart things surrounding us could recognize our desires, requirements, and interests and act accordingly without intervention by any external instructions [3], [4] from any other human or machine. Following Fig. 1 shows how people things and cloud is at edged of IoT triangle.

We are the part of society which has been characterized as heterogeneous, dynamic, and of complex nature but also has features to establish social relationships. This characteristic of society encourages us to be a part of community based on several factors such as common interests, influence, needs, etc. We meet, interact, collaborate, and share ideas, information, feeling, and many more among the different members of the community. This way of communication between us may solve many complex problems. Social technologies help to increase this collaboration between people, permitting the communities to explore innovative ways and methods to share information and ideas and engage the members of community efficiently. Though apparently from different domain, in this era of technology convergence, it appears to have growing synergies between IoT and social technologies [12].

This concept of social networks can easily be incorporated in IoT. The prime motive of IoT devices are connectivity between environment through physical smart device or “Things” to network while people use social technologies to collaborate and interact socially. Adopting these social features for the IoT paradigm has given birth to a new concept of social network of smart objects, services or both, and named as Social Internet of Things (SIoT), that definitely suffice the needs of users, developers, and designers [5–8].

This phenomenon will further boost the collaboration capabilities of IoT devices to deliver information about human–computer interactions with very limited interventions from humans or some time without human interventions. With the human-like capabilities and social interactions between the devices a high degree of interface can be achieved. SIoT definitely boosts the information shared, supporting new innovative applications and definitely provides reliability and a trustworthy network connectivity of social objects. The SIoT is not a single, unified network of socially connected devices, but is a set of number of technologies which can be cohesively work in coordination for the different applications and services and provides an ultimate help for people in both developed and developing economies. In developing countries, SIoT can act as a catalyst in their development by improving the living standards, ease of life, facilities, and infrastructural benefits. It will not only help them grow economically but will also make a significant contribution toward their social and economic development. The aim of this paper is to provide a comprehensive view on the fundamentals of SIoT domain, challenges of SIoT, current scenario of IoT in developing countries, and requirement for implementation for SIoT application in developing countries, future application for the developing countries, and effect of future application to humans.

2 Fundamentals of SIoT

The IoT is outlined by everyday objects, featured as Any time ($24 \times 7 \times 365$), Anything: individuals, objects, data, programs Any Where: on the move, indoors, outdoors, urban areas, rural areas, human unattended areas Any One: old/young, babies: born/unborn, handicapped/healthy, illiterate/literate, male/female Any Service: pervasive service, explicit service, remote service, native service and Any network: multi-protocol, multi technologies, multi-OS. As expressed by a survey conducted by IoT Analytics GmbH in August 2018, the amount of IoT devices are 7B that range doesn’t include smartphones, tablets, laptops or mounted line phones [9]. According to the safe atlast [10] the number of IoT device was 26.66 billion, which were active at the start of 2019 and also the variety exploded to 75B until 2025 [10]. According to statistics shown on Internet, an approximation of 127 new devices added to

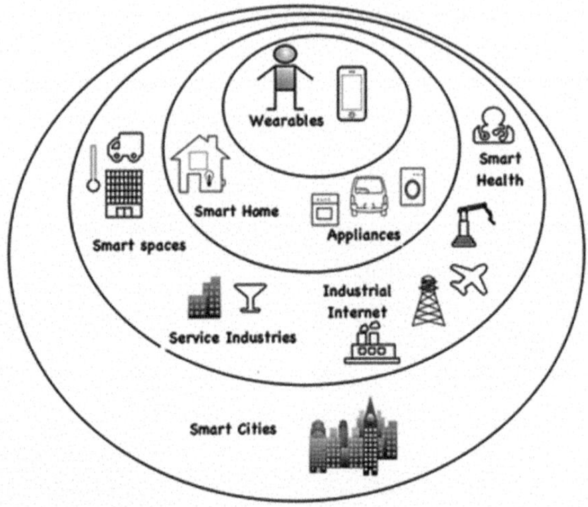

Fig. 2 Existence of IoT in our lives

network every second. There are necessary four dimensions for IoT devices are: identification, sensing, and feeling through networked-sensors; embedding, "things" into systems and investment nanotechnology that modify small things to be part for the environment. Through a gateway, that is actually a collector, these things hook up with the Internet and supply data from the perceived atmosphere to the network and onward to the IoT processors that add up out of the huge quantity of information collected. These processors make intelligent choices and facilitate gives top quality services to users. Over time, we tend to expect nearly everything, possibly, to attach to the network permitting pursuit (location), and watching (status of the objects) in real time to a complicated set of processors within the IoT system. As the number of devices increased day by day, the existence of IoT device in our surrounding is also increased. Furthermore, we have numerous options for selections to pick applications or services in current scenario and therefore interaction to IoT devices is extravagant nowadays. The following Fig. 2 shows the existence of IoT devices in our lives.

The cyber-world creates a world for a user without borders of country, language, color, and creed. The social media provides a platform to share the user details, their habits through their posts, tweets, etc. According to records on an average 6000 tweets done by the users on Twitter every second which is around 350,000 per minutes, 500000000 per day, and 200 billion tweets per year [29]. According to report published in [30] out of 100 user 51 users access the platform every day and around 31 people accepted that they use it several times in a day. 8.95 million photos and videos are shared on Instagram per day [30].

In the order of same videos on Facebook were watched for 100 million hours. The average time of users spent on Facebook is increased to 35 min from 20 min. Largely used Facebook 400 new registration as user, 317,000 post were posted; 54000 link were shared by user, 147,000 photos uploaded every minute [31]. As status show that SIoT is also growing exponentially in society. Following Fig. 3 shows that SIoT

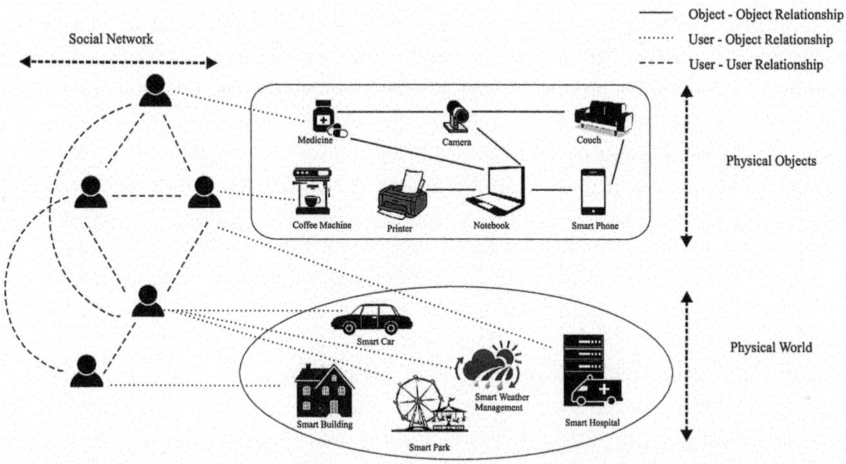

Fig. 3 Components of SIoT [12]

consists of IoT and social network. Communication between the connected objects is increased exponentially and can be exploited to analyze the social behavior of the users of such devices.

This is given a new method of identifying the situation and provide a quick response. This can also be used to have some critical situations as well to provide help to the user.

The technologies support IoT are like ontologies matching [11], machine learning [12], common sense reasoning [13], wireless Communication [16], deep learning [14, 15], and human–computer interfaces [16], among the various are progressively contributive to the event of SIoT.

3 Challenges in SIoT

The number of connected devices in 2019 is more than twice the size of the world population. These hackers will simply breach a sensible device and spy on the house owners. They'll get into industrial management systems, manipulate production lines, shut down factories, and even cause shipwrecks. An easy hack will flip your child's sensible toy into a spying device. Internet-connected medical devices are often compromized and cause someone's death. The majority of IoT machines have some security problems. Same as this there are several challenges related to the IoT. We cannot evade the historic period and therefore the market is ready to succeed in $16.86 billion by 2025, in keeping with Grand view analysis. With a large corroborative infrastructure developed countries still facing issues in implementation. Challenges are in dimensions from technical, designing, managing, structure, sociopolitical extended to the business of recent technology that guarantees to vary the means we tend to live and add major ways. Different literature describes different challenges in keeping with their criteria. Due to all above reasons, IoT businesses are not very successful. Around 75 businesses out of 100, newly opened business either suffer from losses or shut down according to the reports of Internet of Things statistics, 2019. The main reason for the failure is stated as lack of understanding of the data in the IoT systems. Another common reason for failure is limited internal expertise. A high involvement of multi-disciplinary experts as well as domain experts is required to make use of data which is generated by the IoT systems. Internet of Things statistics reports stated that decoding the data for the application, use of that data, measures for quality of data, lack of experts, methods for makeup for budget overruns, are the major managerial reasons behind the less success for these IoT projects. In keeping with Oxford web Institute [17] mentioned some Social, legal, and moral problems regarding IoT are: Privacy and data protection; Global misinformation systems; Big data problems; Public attitudes, opinions and behavior; Tightly coupled systems; Quality of service issues; New forms of risk; Linking the SIoT to work on responsible innovation. Some of the technical challenges,other than above-stated challenges are as follows:

- Scalability: In form of Sensors and actuators IoT system has a number of smart device to work together. Management of the large number of smart device which are connected to SIoT should be taken care.
- Dynamicity: As the number of devices are mobile, and joining and leaving the network very frequently, the system should be able to manage the change topological changes in the network.
- Mobility: In some of the applications devices in the system are mobile. The system should manage the change of location in case of mobile devices.
- Heterogeneity: IoT systems are development for the specific application area. IoT systems are not general purpose in some way. Devices used for an application are meant for the specific application. An application also have number of different types of device and interaction methods. The management should be concerned about the heterogeneity.
- Opportunistic Existence: As the component of IoT system is highly heterogeneous and may be mobile, the system should manage status of this dynamic interaction with the system or outside the system.
- Interoperability: with the diversity in the components, the management system should function with different types of data source. The system should be able to manage this at different data levels also.
- Survivability: IoT system communicate with different components and network in different ways. Trust management protocols must be designed to protect from those malicious attacks. The system defines a secure communication channel as in most of application IoT systems are used wireless communication which are considered more vulnerable than dedicated wired communication.

- Adaptability: Each IoT system interact with the environment in one or other way. Environment is characterized by changes. The IoT system should be able to adapt all the environment for the application.
- Standardization: Till now IoT systems are used with hit and fail methods. They are application specific and use local systems. A worldwide standard system does not exist for the same. The community should be realized with widely accepted standards and are required to make system sustainable.

In addition to the above challenges, developing countries are also facing some other issues. Infrastructure could be a crucial challenge for developing countries. IoT applications must possess the next infrastructure. Internet connectivity in developing countries is very less (34.1%) as compared with the global average (95%). Besides, low budget, unstable political environment, a dearth of investment, an education system that does not support innovation, unemployment that leads to migratory populations, a large part of population that has no proof of identity, people who are stuck fulfilling the basic necessities of life, with no time to think about the big picture, and other such issues, are factors that weave the landscape across developing countries.

All of this makes the adoption of IoT as a big challenge. Security problems are vital while implementing IoT safely. This challenge is vital because of the billions of devices connected through IoT, it needs an economical security mechanism. IoT uses all reasonable information and this ought to be protected. This issue is more problematic for developing countries because of the vulnerable systems [19]. Considering the connected devices which is simply copied, where privacy problems are another challenge to overcome [20]. Supporting atmosphere and infrastructure as power, poor pollution, extreme temperatures, high levels of humidity, and also dirt are poor medium coverage turning into a hurdle for development.

4 SIoT in Developing Countries: Today

Connectivity brings way to several advantages to neglect challenges. Therefore, IoT is presently become one thing as natural as electricity or water. In everyday life we can simply think that by how many number of IoT devices we are surrounded with. You've got fitness trackers sending data to smartphones or sensors measure pollution in the cars that run with the assistance of an application. Africa, uses IoT which is used for public safety, IoT for water quality observance, in Republic of Benin and Gold Coast IoT is employed for pollution observance, in Kenya, Barma they use IoT for weather observance, China is using IoT for pollution observance, and Ruanda uses IoT for tea plant management. Here we can find, developing countries like China is troubled to manage their pollution levels. Africa has long been awaiting for clean water technology, and India, that contains a staggering population of one billion, has been troubled to supply employment to its citizens.

For instance, a high population means that the generation of additional knowledge is more in analysis and innovation. Moreover, issues faced by these countries will open up numerous and unknown fields, like the property atmosphere, for the appliance of IoT. IoT as a tool is helpful in every domain and share of investment in IoT systems is increased in last years which will continue for the future [26]. However, to use them in environmental applications like water management, atmosphere control, healthcare, soil health, structural health, agriculture, emergency services, waste management, safety, etc. for the developing countries will help to justify the requirements but is not economically sound and affordable to a billion population is incredibly tough. Government initiatives, supporting atmosphere, sensible living standards try to increase approvals for good applications which play important roles within the market. A numerous example can be listed with IoT device. But the stream-less work can only be achieved after we deal with various challenges.

5 SIoT in Developing Countries: Future

IoT trends demonstrate the market share for IoT systems are going to be increased exponentially as to $1.3 trillion by 2020, which is almost double from 2016. According to business trends the prediction for annual growth of 15.6% is anticipated. In the current scenario the 54 of successful business acknowledge their success to the use of information technology and IoT collaboration. Also 49 other business leaders agree to encourage the IoT project with the monitory or other support.

A bis number of business firms are using the IoT systems for different tasks and they also want some solution to their problems from the domain. The industry not only works for businesses, but the efforts made is also need to be predicted for the smart home and city or country governing bodies. This tends for an assumption of selling of 1.9 billion sensible home devices which are going to be sold by the year 2019. Sensible home devices or personalize devices change our lives drastically. They are becoming essential from step counter to heartbeat tracker. We are using many of the devices to make our lives convenient.

Due to this change, business firm also encourages to invest in the IoT system, expecting to urge a cut of the $490 billion profit created on the nearly two billion devices that may be sold-out by the tip of the year. The number of wearable devices is also predicted to increase by 191.5% which is more than sale increment in any other business. With very small size and affordable pricing, these devices anticipated more than this. 82.5 million wearable devices are expected to be sold till 2020. These devices are also essential in many SIoT system as they are first point of connection to humans. As in future SIoT infrastructure can be used for home to security, from Transportation to logistics, from agriculture to health, and from smart home to smart cities [26]. Figure 4 shows the potential use of SIoT in developing countries.

Following are some future application can be used in developing countries with SIoT

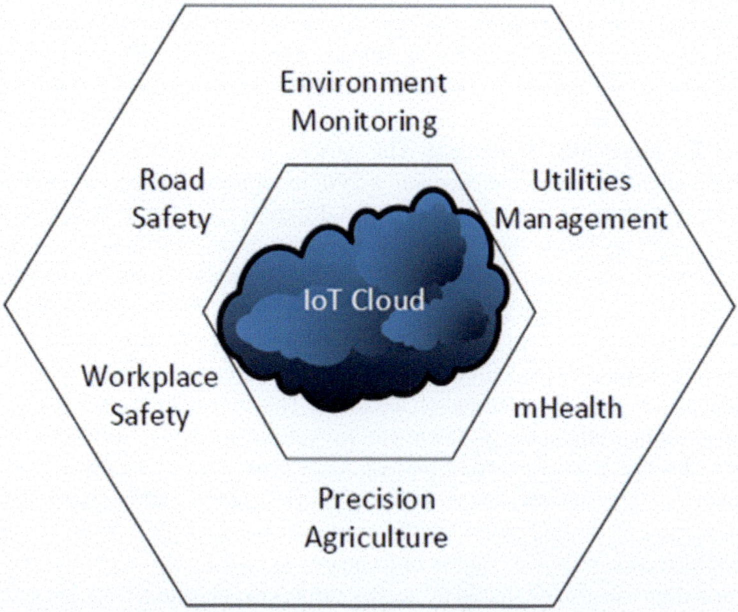

Fig. 4 Potential use of SIoT in developing countries

1. **SIoT for agriculture:** Economic of developing countries are based on manufacturing agriculture. Major problems faced by developing countries are scarcity of food and employment. SIoT can be used for 'precision farming' where with lesser resources to produce more and more farm production. This is the era of drones. Imagine the scenario: A farmer enclosed his land in a glass cabin, harvest is monitored by quadcopters, drones, and driverless tractor, which are there to provide detailed information about the soil, water content, nutrients in the soil, and also measuring the growth and progress of every plant, down to the individual vegetable or fruit. IoT helps in the plantation by providing a scale of humidity and temperature to make conditions favorable for the breeding [27]. SIoT devices can also predict or inform sudden change in climate. SIoT also be used for famines and droughts, floods, and large-scale infestation of pests.
2. **SIoT for smart home:** Used automatically in the appliances used for householdworks are some examples of smart home applications [28].
3. **SIoT for child, women, and aged security:** wearable devices can be used for child trafficking, child security and women security. As reported, crime against children and women in developing country is at higher rate. The devices can inform the danger by a simple interface to social media or to a nearer device.
4. **SIoT for health care:** Sensed data obtained from the sensor with a patient, wrong drug, quantity, and timing of drugs can be detected, can be shared to the doctor or closest healthcare organization. This information can be used to help in critical cases or to avoid causalities with appropriate action on time [30]. For

an example, life of 22-year old, Jorge Cox was saved by an alert from a wearable device "Apple Watch" which was showing heart rate of 130 beats per minute with 60 and 100 bpm. With further investigation revealed that he was born with leaky heart valve.

5. **SIoT for physically challenged:** The sensors are very much helpful for physically challenged. The Blind Navigation System supports visually impaired people to identify products in supermarket from different predefined areas called cells and sensor tags locate the item and these tags pin into navigation system [28]. These devices can also be used for the instant help during life-threatening situation which is also the part of IoT that transmit the information to the nearest police station as well as to the nearest hospital in case of emergency.
6. **Smart transportation:** IoT contribute a much for transportation by using sensors and actuators with vehicles and on the roads. This may help in emergency events like accident, vehicle break down, etc. as well as in criminal offences done on the highways or low-crowded places. Also, sensor technologies benefit from camera and pressure sensors in order to find and control parking space that help less trained or disable people to be more independent [29]. SIoT can inform the information to police station or to nearby hospital when required. Accident information can also be shared by nearby vehicle to get instant help. For example, car manufacturer company Ford occupied the car with feature to SOS number in case of emergency.
7. **SIoT in Industry:** One of the largest contributors to the national economy is manufacturing. Of all the points in an operational value chain that IoT can make a difference, including investments in IoT hardware from sensors inserted in manufacturing equipment and products to electronically items with tags in the extended supply chain, the major gain comes when IoT data helps to setup machine learning protocols and helps take decisions in optimizing operations. Improvement of workflows and shift reorganization dynamically are of two possibilities. Sensor data is sued to predict when the equipment repair and maintenance cost is reduced by half and also does reduce the equipment downtime in equal measures. Sensors measuring stock levels now help 'just-in-time' supply chains and autonomous on-the-shop-floor vehicles raise productivity significantly, due to precise evaluation and near-real time monitor machine. Sensors are very much used in industry in one or another way. Self-organizing logistics can make a drastic change in the future and support business task. Real-time response of task in the business increases the productive and reduces the losses in one or another way.
8. **Smart Cities:** We know that there is a massive change in demographics with a signal shift in population from villages to cities. This shift is accompanied by increased stress on existing infrastructure and services, and through Smart Cities initiatives, we expect to see several areas where IoT will provide the foundations for emerging service frameworks. The solutions are many. Investment in creating the right infrastructure could be one. A noteworthy point to make here is that it is also the duty of the developed world to ensure that no country lags behind in the race to development. Smart cities equipped with smart transportation, waste

management, smart hospital management, smart governance, smart light, and many other things.

9. **SIoT for natural resource:** Water scarcity is also a major problem. The world population is now expected to touch 10 billion by 2050 and water scarcity emerges as the singular threat to human existence. With the importance of water for both human and economic development in societies and its acute scarcity in many places, networks of water-sensitive sensors, connected together with the appropriate simulation workflows can help monitor water interventions such as water shed management and catchment area management.
10. **SIoT for Environment Monitoring:** SIoT can be used for Environment Monitoring. They can be used to predict the environment and inform for the same. SIoT system can also inform the weather change in one geographical area to other for better management. They can also sense and inform the natural disaster for better management by governments for human safety. Given the impact of climate change, and increasing cases of nature's fury and disasters, by means of sensors placed in notified and disaster prone areas, and through simulation, disaster management IoT applications can reasonably model and predict occurrences of landslides and avalanches for citizens in those areas to take proactive actions. SIoT gives the new goals to technology as well as making our lives easy. These devices have become essential part of our lives to increase the support system.

6 Conclusion

Internet of Things (IoT) has acquired an extraordinary impact in the last decade. This phenomenal technology is moving toward to design a world where billions of smart, interacting things are able to offer wide range of facilities and handiness with ease of access. Services without interleaving our lives, where Smart products are intelligent items and they are able to get information and communicate with other to make a safer environment. IoT systems, undeniably benefit the society, however, there are some challenges to apply IoT. Speed and process for implementing IoT differs around the world but developing countries have initiated the process and identifying that IoT system may give a new direction to the development and will definitely provide a much safer and secure environment for the world.

References

1. D. Giusto, A. Iera, G. Morabito, L. Atzori, *The Internet of Things* (Springer, New York, 2010)
2. M. Presser, A. Gluhak, *The Internet of Things: Connecting the Real World with the Digital World* (Czech Republic, EURESCOM, 2009)
3. M. Botterman, Internet of things: an early reality of the future internet, in *Report of the Internet of Things Workshop*, Prague, Czech Republic (2009)

4. A. Dunkels, J.P. Vasseur, IP for Smart Objects, Canada (2008)
5. L. Atzori, A. Iera, G. Morabito, Siot: giving a social structure to the internet of things. IEEE Commun. Lett., United States, pp. 1193–1195 (2011)
6. R. Girau, S. Martis, L. Atzori, Lysis, From smart objects to social objects the next evolutionary step of internet of things. IEEE Internet Things J., United States (2017)
7. M. Nitti, L. Atzori, I.P. Cvijikj, Friendship selection in the social internet of things: challenges and possible strategies. IEEE Internet Things J., pp. 240–247 (2015)
8. I.-R. Chen, F. Bao, J. Guo, TrustBased service management for social internet of things systems. IEEE Trans. Dependable Secur. Comput., United States, pp. 684–696 (2016)
9. SurveyEngine GmbH. Potsdamer Platz, Berlin, Germany
10. https://safeatlast.co/blog/iot-statistics/
11. M.S. Gyrard, G.A. Atemezing, Semantic Web methodologies, best practices and ontology engineering applied to internet of things, in *IEEE 2nd World Forum on Internet of Things (WF-IoT)* (IEEE, United States, 2015), pp. 412–417
12. M. Gyrard, J.B. Serrano, S.K.Datta Jares, M.I. Ali, Sensor-based Linked Open Rules (S-LOR): an automated rule discovery approach for IoT applications and its use in smart cities, in *Proceedings of the 26th International Conference on World Wide Web Companion* (Perth, Australia, 2017), pp. 1153–1159
13. M.H.M. Noor, Z. Salcic, I. Kevin, K. Wang, Enhancing ontological reasoning with uncertainty handling for activity recognition. *Knowledge-Based Systems* (Elsveir, Amsterdam, 2016), pp. 47–60
14. C. Otte, Safe and interpretable machine learning: a methodological review. *Computational Intelligence in Intelligent Data Analysis* (Springer, Berlin, 2013), pp. 111–122
15. J.R. Lloyd, D. Duvenaud, R. Grosse, J.B. Tenenbaum and Z. Ghahramani, (2014) Automatic construction and natural-language description of nonparametric regression models, in *Proceedings of the Twenty-Eighth AAAI Conference on Artificial Intelligence, IEEE Journal on Selected Areas in Communications, United States*, pp. 1242–1250
16. Z. Yan, P. Zhang, A.V. Vasilakos, A survey on trust management for internet of things. J. Netw. Comput. Appl., Elsevier, Netherlands, pp. 1242–1250 (2014)
17. J.E. Kim, X. Fan, D. Mosse, Empowering end users for social internet of things, in *International Conference on Internet-of-things Design and Implementation*, Pitsburgh, PA, USA (2017)
18. Z. Yan, P. Zhang, A.V. Vasilakos, A survey on trust management for internet of things. J. Netw. Comput. Appl., Elsevier, Netherlands, pp. 1242–1250 (2014)
19. J.E. Kim, X. Fan, D. Mosse, Empowering end users for social internet of things, in *International Conference on Internet-of things Design and Implementation*, Pitsburgh, PA, USA (2017)
20. 6LoWPAN: Incorporating IEEE 802.15.4 Into the IP Architecture, Internet Protocol for Smart Objects (IPSO), Canada (2009)
21. N. Gershenfeld, R. Krikorian, D. Cohen, *The Internet of Things* (Scientific American, United States, 2004), pp. 76–81
22. I. Toma, E. Simperl, G. Hench, A joint roadmap for semantic technologies and the internet of things, in *Proceedings of the 3rd STI Roadmapping Workshop*, Crete, Greece (2009)
23. A. Katasonov, O. Kaykova, O. Khriyenko, S. Nikitin, V. Terziyan, Smart semantic middleware for the internet of things, in *Proceedings of the 5th International Conference on Informatics in Control, Automation and Robotics*, Funchal, Madeira, Portugal (2008)
24. L. Srivastava, Pervasive, ambient, ubiquitous: the magic of radio, in *European Commission Conference From RFID to the Internet of Things*, Bruxelles, Belgium (2006)
25. G. Marrocco, C. Occhiuzzi, F. Amato, Sensor-oriented passive RFID, in *Proceedings of TIWDC 2009*, Pula, Italy (2009)
26. S. Duquennoy, G. Grimaud, J.-J. Vandewalle, The web of things: interconnecting devices with high usability and performance, in *Proceedings of ICESS'09*, HangZhou, Zhejiang, China (2009)
27. A.H. Ngu, M. Gutierrez et al., loT Middleware: a survey on issues and enabling technologies. IEEE Internet Things J., United States (2017)

28. R. Khan, S. U. Khan, R. Zaheer, S. Khan, Future internet: the internet of things architecture possible applications and key challenges, in *Proceedings of 10th International Conference on FIT*, Vietnam (2012), pp. 257–260
29. https://www.internetlivestats.com/twitter-statistics/
30. https://www.wordstream.com/blog/ws/2017/04/20/instagram-statistics
31. https://www.omnicoreagency.com/facebook-statistics/

Prevention of Phishing Attack in Internet-of-Things based Cyber-Physical Human System

Alekha Kumar Mishra, Asis Kumar Tripathy, Sowmya Saraswathi, and Meenakshi Das

Abstract In Internet-of-Things enabled Cyber-Physical Human System (CPHS), the controller control the destination systems. The challenger or hacker can perform a number of attacks on this network to threaten the identity and vulnerability of the system, by consuming the networked resources. One of the issue that possesses threat on identities and user credentials is the phishing. The mechanisms for phishing detection in IoT based CPHS should be light-weight and not much complicated in order to meet the CPHS requirement. In CPHS, the credentials can be compromised from the user by showing very similar electronic pages or messages, and encouraging user to provide their secured financial data. These issues need to be resolved in order to get the right output and get all the functionalities to work properly. CPHS has mainly two major components, the first one is controller and second one is destination system. Commands are sent from the sensor to the destination via sensor nodes on the network and the destination system communicates with the controller about what actions to perform or how to deal with the information that controller has sent.

A. K. Mishra
Department of Computer Applications, National Institute of Technology Jamshedpur, Jamshedpur, India
e-mail: alekha.ca@nitjsr.ac.in

A. K. Tripathy (✉)
School of Information Technology and Engineering, VIT Vellore, Vellore, India
e-mail: asistripathy@vit.ac.in

S. Saraswathi · M. Das
School of Computer Science and Engineering, VIT Vellore, Vellore, India

A. Nanda and N. Chaurasia (eds.), *High Performance Vision Intelligence*, Studies in Computational Intelligence 913,
https://doi.org/10.1007/978-981-15-6844-2_2

1 Introduction

Phishing attack is one of the cyber threat in IoT based Cyber-Physical Human System that is able to compromise the user credentials and confidential information with a set of easy and simple steps. Phishing attack can be addressed in four generic ways based of the complexity of the approach. Simplest way is to literate the users about phishing attack and the precautions to be followed through different awareness and education programs [1, 2]. However, this approach is ineffective in the current level of phishing attack because it targets the vulnerability in the user access methods. The second way is to implement preventive measures to protect system from getting penetrated by phishing attack. This is commonly achieved via end-to-end authentication schemes [3]. This approach is more expensive and not adaptive to handle the trust forging ability of phishing attack. Third way is to detect phishing attack using various approaches to differentiate a genuine website from phishing website [4]. This approach can address phishing attack efficiently due to provision of using most effective techniques such as machine learning to generate and incorporate filtering rules to detect phishing. Finally, the fourth way is to take legal action and bring down the phishing websites. This approach is too expensive and tedious in the current scenario of the Web browsing.

The content of the chapter is arranged in the following order: Sect. 2 provides fundamental observations made about phishing attack and its procedure. Section 3 classifies the phishing detection techniques based on the features and approaches as reported in the literature. Section 4 discusses detection techniques by systematic representation of detection procedure for each of them. An analysis and comparison of each detection technique is provided in the Sect. 5. Finally, concluding notes and future plans are briefed in Sect. 6.

2 Phishing Attack

The phishing attack [5] is often launched with an objective to steal user credentials related to their financial data by mimicking legitimate electronic communications. The target environment of a phishing attacker is mostly e-commerce, banking, and informative sites. The commonly stolen credentials include user login-id and password, credit-card number, ATM pin, etc. As a result of this, a victim suffers huge financial loss and defamation [6].

The most common form is the email phishing, also known as deceptive phishing attack [7]. In this form of attack, an attacker sends thousands of spam mails. Even a single successful attack can result in a lump sum amount of confidential information. The phishing mail mimic the actual emails using same phrasing, typefaces, logos, and signatures to make the message appear legitimate. They usually push user into action by creating a sense of urgency using phrases like "expiry date of card, password, or membership". The link inside email visually closer to original site, but typically

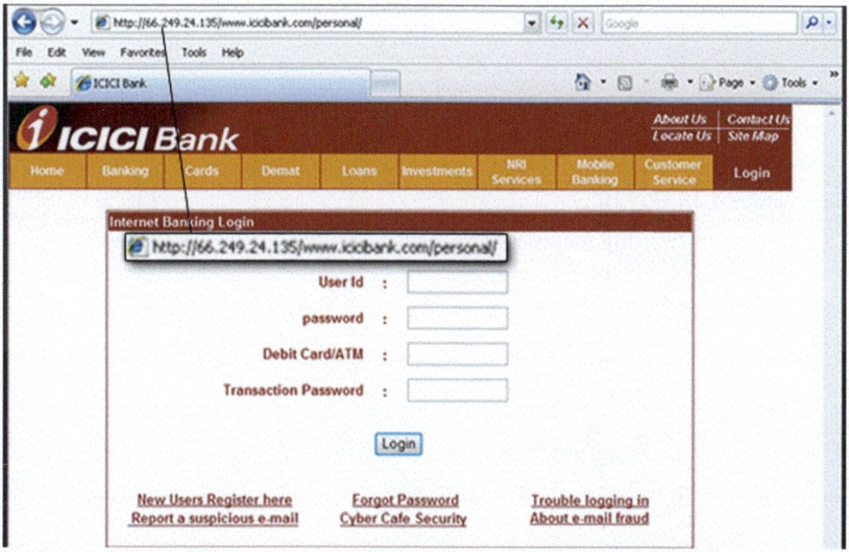

Fig. 1 An example of phishing page for ICICI online banking page.{source: google}

represents a domain name or extra sub-domains with one or more misspelled letters in it. An example of phishing mail has been shown in Fig. 1.

Spear phishing is a deeper and sophisticated form that targets a specific person or enterprise [8, 9]. It requires detailed knowledge of an organization including its privileged employee hierarchy. Knowing a particular person's username and password at administrative level can help to launch attack on the entire organization. Spear phishing launched at the higher level of the organization's employee hierarchy is also called whaling. The first approach verifies the authenticity of the host name, url, images, logos, etc. in the pages using a learned heuristic based on the past history. Basically, this approach looks for a pattern it has already learned. The second method, however, is more like a look-up list method. This method looks for a list of URLs that are blacklisted as phishing sites[10]. Any link found to be in this list is classified as phishing attack. Some technique uses the combination of both the above two approaches.

In Fig. 2, a classification of the phishing detection techniques as reported in the recent years has been presented.

3 Classification of Phishing Detection Techniques

The phishing detection techniques are broadly classified into the following categories:

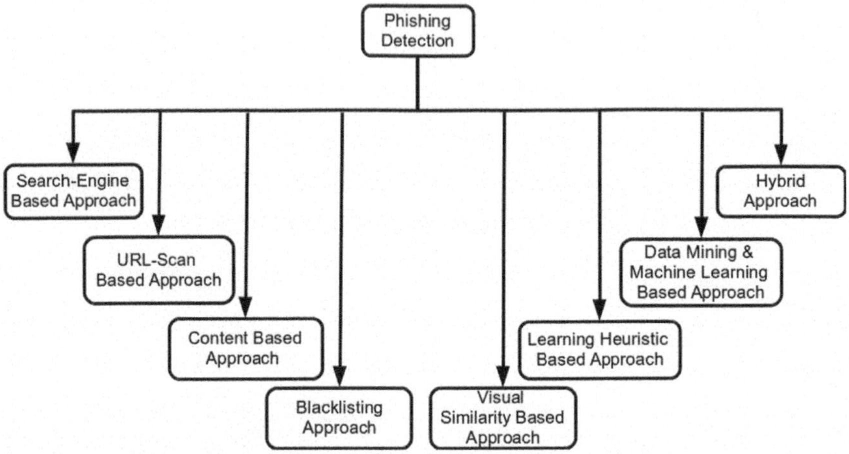

Fig. 2 Classification of phishing detection techniques

- *Search Engine-based techniques*: This technique uses search engines to search for legitimate websites using various parameters such as URL, page content, images, etc. in order to identify the phishing websites. However, the search engine has no role in deciding whether a site is phishing or legitimate. In other words, search engine aids the algorithm with legitimate websites URL and content to evaluate and compare the legitimacy of the requested page.
- *URL-scan-based techniques*: In this category, the proposed systems scan the suspicious URL to parse various elements of interest in order to study and analyze their pattern to identify the anomaly and detect phishing sites.
- *Content-based techniques*: Unlike, URL-scan, these techniques process the requested email content and/or URL's complete webpages, elements, DOM object, etc. to identify keywords and pattern to detect phishing pages.
- *Blacklisting techniques*: These techniques maintain and update a list of already detected phishing sites/links, URLs, and domains. A newly arriving link or URL is looked up for a closer match in the blacklisted sites. If a match is found with higher similarity ratio then it is marked as phishing site and is updated in the list of blacklisting sites.
- *Visual similarity-based techniques*: These techniques focus on the elements in a webpage that contribute to the visual appearance of a site. These elements include logos, images, font size and type, alignment, text location, etc. The aim is to find the visual similarity between a reported site and a corresponding popular/legitimate site. Phishing is detected based on the score of similarity between these sites.
- *Learned heuristic-based techniques*: These technique relies on a set of anomaly patterns that may be found on a phishing site. The identification of these anomalies is established through a list of heuristics that is generated through various phases of observation. If a reported website possesses one or more anomaly pattern, then it is detected using established heuristic.

- *Hybrid techniques*: These techniques combine one or more abovementioned techniques or with other detection techniques to improve the efficiency. Machine learning-based hybrid technique are more popular due to the ability of inferring new possible patterns of phishing from the existing ones.

The abovementioned classification is pictorially depicted in Fig. 2. The following sections provide a detailed discussion of detection techniques of each category as listed above.

4 An Overview of Popular Phishing Detection Techniques

4.1 SpoofGuard

Chou et al. [11] have found that a phishing website shows abnormalities in Domain Object Model (DOM) elements and HTTP transactions. The abnormalities may be used in addition to an existing technique to achieve higher detection rate. It is a web-browser plug-in that detects HTML phishing pages by measuring the changes found in HTML content. Once example of such anomaly is <a href="randomphishing.com" > facebook.com </a>.

4.1.1 Detection Procedure

First, DOM objects anomaly heuristics are defined. These heuristics-based rules are then incorporated as a part of SpoofGuard and the event is defined to be executed when page is getting loaded in the browser. Each object present in a page is checked if the objects structure and properties satisfies any of the defined rules in SpoofGuard. If found, then it is detected as a phishing page.

4.2 Antiphish

It is a browser plug-in proposed by Raffetseder et al. [12] that keeps track of the paths of sensitive information of an user such as login credentials, card number, etc. This path includes the connections to the website, where user regularly performs online operations and financial transactions. The plug-in uses these information to alert the user when the same credentials are transferred in a different path than the recorded (trusted) ones. However, Antiphish also marks the genuine credentials as suspicious when used in paths other than the normal path.

4.2.1 Detection Procedure

Antiphish tracks and records the path and destination of all out going sensitive information. If the sensitive information flow is tracked in a path and the website that is other than the previously entered trusted site, then generates an alert message to the user using a high alert pop-up message.

4.3 *Natural Language Processing-Based Anti-phishing Technique*

Stone has proposed a detection system and named as EBIDS [13]. EBIDS is a Natural Language Processing (NLP)-based anti-phishing technique that makes it unique from other techniques despite of its average performance accuracy. It is reported that EBIDS has false positive detection rate of 1.9%, and false negative detection rate of 25%. These figures are not competing compared to its counterparts.

4.3.1 Detection Procedure

The first task involves the extraction of message bodies and sending it to the detection procedure. The mechanism uses ontosem to classify the message. The signature rules are defined using string literal matching against ontosem's output. Depending on this result, an email is reported as legitimate or phishing. The financial opportunity rule detects the emails on financial opportunities, such as heavy discount and free travels and tickets.

4.4 *Phishing Sites Blacklist Generator*

Sarifi and Siadati [14] have provided a simple yet efficient procedure to generate blacklisted websites under phising attack. The proposed technique is based on the fact that phishing pages use legitimate site's logo to make their page appear genuine and part of legitimate website and most of the sites show their relations using logo. Therefore, the logo can be used to determine the page which belongs to legitimate website. This algorithm is more useful when we use it for emails to generate blacklist of phishing sites.

4.4.1 Detection Procedure

First, datasets of legitimate and phishing pages are fed to the generator. Then, the organization name is extracted from the suspected URL and searched in a search engine to return top ten results. If the suspected URL is from the top ten results, then site is assumed to be legitimate. Otherwise, the site is detected as phishing site and recorded in the blacklist.

4.5 *Phishwish*

Cook et al. have proposed Phishwish [15] that provides eleven important rules to analyze the URLs provided within body of an email for detecting phishing. It claimed to be resilient against zero-hour attacks other than blacklists. The technique also has lower false positives detection rate.

4.5.1 Detection Procedure

Incorporate the following rules and infer phishing if one or more following rules hold true for any of the URL in the email body.

1. The email is formatted as HTML and URLs with the page uses Transport Layer Security (TLS), but HREF attributes do not use TLS.
2. Host name is an IP address.
3. URL path mentions an organization name, but not in the domain name.
4. Inconsistency in Whois records.
5. Inconsistency in image's Whois record.
6. Page is inaccessible.

4.6 *Phishguard*

Joshi et al. have proposed Phishguard [16] that acts as an intermediary between the user and the website. In this technique, HTTP digest authentication is used for detecting phishing. The passwords for the sites are stored instead of URL. Therefore, even though the attacker can access this storage, she cannot identify the matching relevant site of the stored credential. The randomly generated passwords help to protect to easy distinction of correct password from wrong ones through visual or other ways of inspection.

4.6.1 Detection Procedure

When the user visits a login page, he/she enters the credentials. The Phishguard traps the original credential and sends fake credential(s) to the phishing website a random number of times. In this way, Phishguard stores the correct credentials of the user and validates the future login against the phishing website.

4.7 Visual Detection Without the Victim's Site Information

Hara et al. [17] have proposed a visual-based detection technique without using victim's site information. That is, this does not need white-listing of legitimate websites. In addition, this technique also analyzes the header portion of HTTP messages. However, it suffers from higher false positive rate and the use of image processing techniques on client browser may raise a significant delay in the browsing experience.

4.7.1 Detection Procedure

The suspicious URL is visited by the web browser and the images displayed on the browser are extracted. An image database is maintained that consists of pair of URLs and their images for all genuine websites. The URL and extracted images are searched in the database for a match.

4.8 Bogus Biter

Yue et al. have proposed a technique called Bogus biter [18]. It is available as an extension to Firefox 2.0 or higher version web browser. This approach is based on the blacklists provided by Google. It defends against a phishing attack without the help of a third party's service. It is also transparent to user. However, the toolbar is required to be installed in a large number of user's browser to improve its effectiveness. One side effect of this technique is that it may result in Denial of Service (DoS) by flooding against servers hosting legitimate sites but only one of them have phishing data.

4.8.1 Detection Procedure

Initially, URL blacklist is prepared and updated on routine basis. For detection, the selected URL is checked in the blacklist. If URL is blacklisted, then a huge number of fake credentials are fed to the URL. At the same time, the victims actual credentials are concealed with the bogus credentials and legitimate URLs are enabled to identify the concealed credentials in a regular time interval.

4.9 *PhishNet (Predictive Blacklisting)*

Prakash et al. [19] came up with an observation that most of the blacklist bypass techniques are more or less similar URLs with two or more host names may resolve to the same IP address. Authors provided a measure of deviation from the exact match implementation to approximate the match of blacklisted URLs. The proposed PhishNet tool is used to achieve approximate matching. The result shows that PhishNet outperforms all detection techniques with blacklist approach.

4.9.1 Detection Procedure

The maliciousness around the original blacklist is computed and possible variation of each URL is produced using five abovementioned heuristics. The common procedure of checking the blacklisted URL is modified to find an approximate match instead of looking for exact match with the current and predicted blacklist. If match is found with value more that a pre-defined ratio, then it is detected as phishing URL.

4.10 *New Content-Based Filtering Approach for Phishing Emails*

Bergholz et al. [20] proposed a new filtering technique for detecting phishing attack based on some novel features. These features include email topic description model with limited dimension, logo identification process, and hidden salting detection. They have used dynamic Markov chain to characterize the text sequences bit by bit and calculate the probability of a message originated from a particular class. Support Vector Machine (SVM) tool is used to define the classes of emails based on email features. New logo detection and image distortion technique are also used to detect phishing emails.

4.10.1 Detection Procedure

A set of features are extracted by scanning the email content and structural properties. Let the feature set be $x = (x_1, x_2, \ldots, x_m)$. Then, it establishes a function to find the class y of an email (phishing or non-phishing), where $y = f(x, \lambda)$. The vector λ is computed using a training process. During training, the relationship between x and y is reproduced in the training set. SVM classifier is used for this purpose. The emails that are unique in the trained set leads to the higher uncertainty of classification. If class probabilities are other than the value of 0 and 1, then these emails are considered to be borderline cases. Finally, the resulting borderline cases are marked in the inbox and left to the user to manually check these, so that they could be labeled as ham, spam, or phishing and added to the training set.

4.11 Large-Scale Automatic Classification of Phishing Page

Whittaker et al. [21] have proposed a technique based on precision over recall. This technique has been implemented by Google which is able to classify phishing page faster while taking care of false positive. To achieve better results, machine learning system is used. Gmail spam filters receive about five million spam URLs per day. Therefore, it is used to prepare training set. A training dataset of about ten million URLs is used to train the classifier in order to pick up new techniques used by phishers.

4.11.1 Detection Procedure

First, URL page features are extracted from the blacklist and from the spam messages of Gmail. It is taken care that the user identifiable urls are not fetched during this process. In the second step, the domain information about the page is obtained. During this operation, it is checked whether the URL containing the IP address for its host name and whether the page has many host components. The metadata of the page is also extracted including the page rank. In the third step, the page features from the page contents are extracted by content fetcher that resolves the host and records the returned IPs, name servers, and name server IPs. Content fetcher also sends the URL to a pool of headless web browsers to render the page content. The browser renders the page and return the HTML content, iframes ,images, and javascripts embedded in the page back to content fetcher. DNS entry's accuracy is also measured during rendering. Finally, the page is assigned a score based on gathered features representing the probability that the page is phishing. The score 0.0 indicates legitimate page, whereas 1.0 indicates a phishing page. The blacklist aggregator then reads the set of blacklisted urls and transforms them into host/path suffix expressions without overlapping expression, if any, and updates the remaining in binary format for safe browsing of the client.

4.12 DBSCAN Clustering-Based Phishing Detection

Liu et al. [22] have used DBSCAN clustering technique to detect whether a webpage is similar to a popular or branded webpage (legitimate). The similar webpages are identified as the phishing websites for the corresponding legitimate websites and are blacklisted. This technique queries search engines over the Internet to find similar pages with their rank that are indirectly related. Therefore, it has an additional network communication overhead and may be a limiting performance factor under constrained network conditions.

4.12.1 Detection Procedure

Initially, all the associated webpages to the requested webpage, say P are gathered using direct and indirect relationships. In the next step, the strength of each of link, page ranking, text similarity, and layout similarity relationship is calculated and the scores are stored in a vector $V_i = \{L_i, R_i, TS_i, LS_i\}$ for each associated webpage. The Density-Based DBSCAN algorithm is then applied to these vectors until P becomes a core point. In case, P is a border point, no cluster is formed and the process is terminated. At the end of this process, if a cluster is found that includes P, then it is inferred that P is a phishing page and the original page is identified as the one which has a strong association relationship with P.

4.13 DNS-Based Blacklisting

Wardman et al. [23] have proposed a DNS-based blacklisting approach for phishing detection. It maintains a blacklist of URLs at DNS level. This technique is currently in use in almost all email systems. However, it requires a server with higher capacity of handling large number of DNS records. The performance of the technique may degrade when installed with a computer with normal computation capability. Secondly, this techniques can be bypassed by the phishers either by acquiring access to legitimate system or using a different IP address.

4.13.1 Detection Procedure

A Domain Name System Blacklist (DNSBL) server is configured and maintained for blacklisted addresses at an organizational level. This server also acts as spam blocker and filter. Before establishing an SMTP connection, the Mail Transfer Agent (MTA) verifies the connecting source by sending a query to DNSBL. DNSBL replies back the result of lookup into its current blacklist. Based on the reply from DNSBL, connection is either established or rejected by detecting it as phishing attack.

4.14 PhishZoo

PhishZoo [24] was proposed by Afroz and Grandstadt in 2011. According to them, objective of a *phisher* is to make the appearance of the phishing website close to its original website. A simplest and efficient way to use an automated tool to download content of the original website and compare it with the reported site for phishing.

4.14.1 Detection Procedure

The host name of a new URL is tokenized into a set of strings. These strings are delimited by “/”,“?”, “.”, “,”,“=”, “-”, and “)”. The tokenization process is followed by extraction of HTML files from the URL.

4.15 DNS-Poisoning Phishing Attack Detection

Kim and Hun [25] have used the network communication characteristics to differentiate legitimate sites from their DNS-poisoning counterpart sites. The proposed detection technique claims that there is significant deviation in the network parameters of a legitimate site and its corresponding phishing sites. In this technique, the routing information of the URL is extracted using IP datagrams and analyzed to detect phishing attack. The common parameters to detect phishing sites are Round Trip Time (RTT) and hop count of network datagram.

4.15.1 Detection Procedure

The parameters such as RTT and hop count for the datagrams that arrive from the original websites are recorded and updated. A K-Nearest Neighbor (KNN) classifier is trained to learn these parameters from corpus of original sites. The information from the datagram about the URL is extracted and the deviation from the original websites is computed. If the deviation is beyond the limit, then URL is detected as phishing site.

4.16 A Textual and Visual-Based Anti-phishing Using Bayesian Approach

Zhang et al. [26] have provided technique that combines bayesian approach and visual-based approach for text and images of a webpage, respectively, to detect phishing.

4.16.1 Detection Procedure

Visual similarity technique takes snapshots of websites and transforms into images of size 100 x 100 pixels. Upon diagonalizing the images, it returns a normalized number that defines the similarity of the images. The individual probability of visual similarity and textual being correct is calculated to a normalized form that lies in the given subinterval. If the Bayesian technique probability is greater than visual then output of Bayesian is used for class label, otherwise visual probability is used.

4.17 *Logo Similarity-Based Phishing Detection*

Chang et al. [27] have proposed a method to detect phishing using identify comparing based on the similarity of website logo. In their work, the authors use segmented logo of the suspicious website to search the google image database to extract website identity. This identity information is used to compared and mark phishing website.

4.17.1 Detection Procedure

The process is divided into two phases. In Logo extraction phase, the image is segmented to obtain the best fit of the logo from the site. The logo is then used to search google image database to obtain the genuine website that contains this logo. Provided with the original identity and owner of the logo, a phishing website is distinguished from the legitimate website by comparing domain names.

4.18 *NCD-Based Phishing Detection*

Chen et al. [28] have come up with a visual-similarity-based anti-phishing system. The proposed system is able to identity the webpages that are visually accurate matching with a legitimate websites to confuse the common user. The system uses Normalized Compressed Distance (NCD) as a metric to compare the object in the webpages using a compression algorithm.

4.18.1 Detection Procedure

First, the white(legitimate) pages are selected to pair and then compared with the requested page. A byte-stream compressor is used to compute the NCD, which is defined as

$$NCD(x1, y1) = \frac{C(x1, y1) - min\{C(x1), C(y1)\}}{max\{C(x1), C(y1)\}}, \tag{1}$$

where $C(x1, y1)$ is the compressed concatenated string representing x1 and y1, and $C(x1)$ is the compressed string x1. In the next step, sixteen heavily phished legitimate brand from financial industry are cached. In the final step, pairwise NCD between the pairs is computed and the resulting features are passed to classifier to infer whether the requested page is either safe or a phising page.

4.19 LinkGuard

Shekokar et al. have proposed a phishing detection technique and named it as LinkGuard [29]. In addition, it also uses blacklisting approach for phishing detection. DNS name is the key factor to decide a phishing attack. A seed set is maintained containing the DNS names that the user inputs manually and assumed that these names are trustworthy. A similarity index between a DNS names from the seed set and actual DNS name is calculated. If the similarity index is high, it is concluded to be a phishing site.

4.19.1 Detection Procedure

In the first step, DNS names from the actual and visual links are extracted and compared. If the names are different, then the actual link is checked against the blacklist. If an entry is found, then the message "PHISHING" is returned, otherwise "NOT PHISHING" message is returned and visual similarity checking procedure is initiated. If an IP address is used instead of DNS name then "POSSIBLE PHISHING" message is returned and proceeded to visual similarity-based checking. In this process, the actual link and visual link are decoded and the snapshots of the suspicious page and original page are taken. In the next step, RBG channels of the original and the suspicious images are used to calculate the Root Mean Square (RMS) error. If RMS is 0, it is considered as original webpage, and if it is above a certain threshold it is considered phishing site.

4.20 Target Domain-Based Phising Page Detection

Ramesh et al. [30] proposed a scheme that not only detect the phishing page, but also automatically warns the target domain that is victim of the detected phishing website. The proposed scheme utilizes the in-degree link associations to establish relation between the website and its domains, and this value is used to distinguish a phishing page from its target domain. The contribution also includes a novel Target Validation (TVD) algorithm to reduce false positive rate of the proposed system.

4.20.1 Detection Procedure

The detection process starts with the count of distinct domains reachable up to two levels of a suspicious page under investigation. This count helps in determining the Target Domain Set. In the second step, a cost matrix is generated based on domains relationship in the Target Domain Set. The proposed Target Validation

(TVD) algorithm is used to ensure correctness of the target domain. The phishing is detected by checking for match of the validated domain with the confirmed target domain.

5 Analysis

The summary of performance evaluation is presented as claimed by the phishing detection technique in the literature. For this purpose, we selected only two metrics that are sufficient to reflect the efficiency level of a proposed technique. These are True Positive Rate (TPR) and False Positive Rate (FPR), which are defined as follows:

$$\text{True Positive Rate (TPR)} = \frac{\text{Number of sites detected phishing}}{\text{Total number of phishing sites}} \tag{2}$$

$$\text{False Positive Rate (FPR)} = \frac{\text{Number of legitimate sites detected phishing}}{\text{Total number of legitimate sites}} \tag{3}$$

Table 1 summarizes the comparison of claimed TPR and FPR of the presented phishing detection techniques. It is observed from the Table that the higher TPR is achieved mostly by the data mining and machine learning-based techniques. This shows that the ML-based techniques are more suitable when accuracy is the objective of concern. Textual and visual-based anti-phishing [26] and NCD-based phising detection [28] are the few among visual similarity-based technique that have similar performance compared to ML-based technique. However, the data set of NCD-based phishing detection [28] is comparatively smaller than overall ML-based techniques. Though content-based techniques are computationally efficient than ML based but only PhishNet [19] and phishWho [37] have higher TPR. The only proposed technique with 100% TPR are LinkGuard [29] and BogusBiter [18], however these performances are not acceptable for implementation because the number of sites used for evaluation by LinkGuard [29] and BogusBiter [18] are only ten and seventy, respectively. The phishing detection techniques [12, 16, 33] that have not used TPR and FPR as metrics for evaluation.

6 Conclusion and Future Directions

In this chapter, a comprehensive survey of various phishing detection techniques is presented. The phishing attack raises to equally complex and serious level with the advancements in the current scenario of cybersecurity. This chapter has presented a list of phishing detection techniques based on their popularity in the literature. The presented papers are classified based on their approach to handle phishing attack. To detect sophisticated version of phishing, the most of the reported techniques

Table 1 Comparison of claimed TPR, and FPR

Detection technique	TPR (0-1)	FPR(0-1)
CANTINA [31]	0.97	0.1
Antiphish [12]	Not Provided (NP)	NP
NLP based anti-phishing technique [13]	0.75	0.03
Phishing blacklist generator [14]	0.91	0.09
Medvet et al. [32]	0.925+(27samp)	0.074
Phishguard [16]	NP	NP
PhishWish [15]	0.75	0.023
Chen et al. [33]	NP	< 0.001
Hara et al. [17]	0.83	0.18
BogusBiter [18]	1(50phish-20leg sites)	0.001
PhishNet [19]	0.97	0.005
New content-based filtering approach [20]	0.98	0.093
Large-scale automatic classification et al. [21]	0.90	0.01
Lexical Feature-Based Phishing URL Detection [34]	NP	0.45
DBSCAN clustering-based phishing detection [22]	0.91	0.03
DNS-based blacklisting [23]	0.95	0.14
PhishZoo [24]	0.90	high
DNS-poisoning phishing detection [25]	0.99	0.07
CANTINA+ [35]	0.92(unique), 0.98(duplicate)	0.004
Textual and visual-based anti-phishing [26]	0.99	0.01
Logo similarity-based phishing detection [27]	0.75	0.42
Harris-Laplace algorithm-based detection [36]	0.91	0.01
NCD-based phishing detection [28]	0.99	0.01
LinkGuard [29]	1 (10 websites)	0.09
PhishWho [37]	0.99 (accuracy 96)	0.074 (FNR-0.32)
Target domain-based phishing page detection [30]	0.99	0.045

are hybrid in nature. The combination of two or more techniques is intended to achieve higher accuracy. A summary of claimed TPR and FPR is also provided. It is inferred from the table that despite higher computational complexity, data mining and machine learning-based approaches outperform other detection techniques in accuracy and false positive rate.

References

1. T.N. Jagatic, N.A. Johnson, M. Jakobsson, F. Menczer, Social phishing. Commun ACM **50**(10), 94–100 (2007)
2. P. Kumaraguru, Y. Rhee, S. Sheng, S. Hasan, A. Acquisti, L.F. Cranor, J. Hong, Teaching Johnny not to fall for phish. ACM Trans. Internet Technol. (TOIT) **10**(2), 7:1–7:31 (2010)
3. I. Khalil, S. Bagchi, N. Shroff, Analysis and evaluation of Secos, a protocol for energy efficient and secure communication in sensor networks. Ad Hoc Netw. **5**(3), 360–391 (2007)
4. G. Varshney, M. Misra, P.K. Atrey, A survey and classification of web phishing detection schemes. Secur. Commun. Netw. **9**(18), 6266–6284 (2016)
5. J. Hong, The state of phishing attacks. Commun. ACM **55**(1), 74–81 (2012)
6. K.L. Chiew, K.S.C. Yong, C.L. Tan, A survey of phishing attacks: their types, vectors and technical approaches. Expert Syst. Appl. **106**, 1–20 (2018)
7. H. Huang, J. Tan, and L. Liu, Countermeasure techniques for deceptive phishing attack, in *Proceedings of International Conference on New Trends in Information and Service Science*, pp. 636–641 (2009)
8. B. Parmar, Protecting against spear-phishing. Comput. Fraud. Secur. **2012**(1), 8–11 (2012)
9. C. Karlof, U. Shankar, J.D. Tygar, D. Wagner, Dynamic pharming attacks and locked same-origin policies for web browsers, in *Proceedings of the 14th ACM Conference on Computer and Communications Security*, pp. 58–71 (2007)
10. B.B. Gupta, N.A.G. Arachchilage, K.E. Psannis, Defending against phishing attacks: taxonomy of methods, current issues and future directions. Telecommun. Syst. **67**(2), 247–267 (2018)
11. N. Chou, R. Ledesma, Y. Teraguchi, J.C. Mitchell, Client-side defense against web-based identity theft, in *Proceedings of the Network and Distributed System Security Symposium, NDSS 2004*, pp. 01–16 (2004)
12. T. Raffetseder, E. Kirda, C. Kruegel, Building anti-phishing browser plug-ins: an experience report, in *Proceedings of the 3rd International Workshop on Software Engineering for Secure Systems*, pp. 1–6 (2007)
13. A. Stone, Natural-language processing for intrusion detection. Computer **40**(12), 103–105 (2007)
14. M. Sharifi, S.H. Siadati, A phishing sites blacklist generator, in *2008 IEEE/ACS International Conference on Computer Systems and Applications*, pp. 840–843 (2008)
15. D.L. Cook, V.K. Gurbani, M. Daniluk, Phishwish: a simple and stateless phishing filter. Secur. Commun. Netw. **2**(1), 29–43 (2008)
16. Y. Joshi, S. Saklikar, D. Das, S. Saha, PhishGuard: a browser plug-in for protection from phishing, in *2008 2nd International Conference on Internet Multimedia Services Architecture and Applications*, pp. 1–6 (2008)
17. M. Hara, A. Yamada, Y. Miyake, Visual similarity-based phishing detection without victim site information, in *IEEE Symposium on Computational Intelligence in Cyber Security*, pp. 30–36 (2009)
18. C. Yue, H. Wang. BogusBiter: a transparent protection against phishing attacks. ACM Trans. Internet Technol. **10**(2), 6:1–6:31 (2010)
19. P. Prakash, M. Kumar, R.R. Kompella, M. Gupta, PhishNet: predictive blacklisting to detect phishing attacks, in *Proceedings IEEE INFOCOM*, pp. 1 – 5 (2010)

20. A. Bergholz, J. De Beer, S. Glahn, M.-F. Moens, G. Paaß, S. Strobel, New filtering approaches for phishing email. J. Comput. Secur. **18**(1), 7–35 (2010)
21. C. Whittaker, B. Ryner, M. Nazif, Large-scale automatic classification of phishing pages, in *Proceedings of the Network and Distributed System Security Symposium, NDSS 2010*, pp. 01–14 (2010)
22. G. Liu, B. Qiu, L. Wenyin, Automatic detection of phishing target from phishing webpage, in *20th International Conference on Pattern Recognition*, pp. 4153–4156 (2010)
23. B. Wardman, T. Stallings, G. Warner, A. Skjellum, High-performance content-based phishing attack detection, in *Proceedings of eCrime Researchers Summit*, pp. 1–9 (2011)
24. S. Afroz, R. Greenstadt, PhishZoo: detecting phishing websites by looking at them, in *2011 IEEE 5th International Conference on Semantic Computing*, pp. 368–375 (2011)
25. H. Kim, J.H. Huh, Detecting DNS-poisoning-based phishing attacks from their network performance characteristics. Electron. Lett. **47**(11), 656–658 (2011)
26. H. Zhang, G. Liu, T.W.S. Chow, W. Liu, Textual and visual content-based anti-phishing: a bayesian approach. IEEE Trans. Neural Netw. **22**(10), 1532–1546 (2011)
27. E.H. Chang, K.L. Chiew, S.N. Sze, W.K. Tiong, Phishing detection via identification of website identity, in *2013 International Conference on IT Convergence and Security (ICITCS)*, pp. 1–4 (2013)
28. T.-C. Chen, T. Stepan, S. Dick, J. Miller, An anti-phishing system employing diffused information. ACM Trans. Inf. Syst. Secur. **16**(4), 16:1–16:31 (2014)
29. N.M. Shekokar, C. Shah, M. Mahajan, S. Rachh, An ideal approach for detection and prevention of phishing attacks. Procedia Comput. Sci. **49**, 82–91 (2015); *Proceedings of 4th International Conference on Advances in Computing, Communication and Control (ICAC3'15)*
30. G. Ramesh, J. Gupta, P.G. Gamya, Identification of phishing webpages and its target domains by analyzing the feign relationship. J. Inf. Secur. Appl. **35**, 75–84 (2017)
31. Y. Zhang, J.I. Hong, L.F. Cranor, Cantina: a content-based approach to detecting phishing web sites, in *Proceedings of the 16th International Conference on World Wide Web*, pp. 639–648 (2007)
32. E. Medvet, E. Kirda, C. Kruegel, Visual-similarity-based phishing detection, in *Proceedings of the 4th International Conference on Security and Privacy in Communication Netowrks*, pp. 22:1–22:6 (2008)
33. K.-T. Chen, J.-Y. Chen, C.-R. Huang, C.-S. Chen, Fighting phishing with discriminative keypoint features. IEEE Internet Comput. **13**(3), 56–63 (2009)
34. A. Blum, B. Wardman, T. Solorio, G. Warner, Lexical feature based phishing url detection using online learning, in *Proceedings of the 3rd ACM Workshop on Artificial Intelligence and Security*, pp 54–60 (2010)
35. G. Xiang, J. Hong, C.P. Rose, L. Cranor. CANTINA+: a feature-rich machine learning framework for detecting phishing web sites. ACM Trans. Inf. Syst. Secur. (TISSEC) **14**(2), 21:1–21:28 (2011)
36. Y. Zhou, Y. Zhang, J. Xiao, Y. Wang, W. Lin, Visual similarity based anti-phishing with the combination of local and global features, in *2014 IEEE 13th International Conference on Trust, Security and Privacy in Computing and Communications*, pp 189–196 (2014)
37. C.L. Tan, K.L. Chiew, K. Wong, S.N. Sze, PhishWHO: phishing webpage detection via identity keywords extraction and target domain name finder. Decis. Support Syst. **88**, 18–27 (2016)

Blockchain for Secure Internet of Things

Shivani Sharma, Rashmi Chaudhry, and Dinesh Bhardwaj

Abstract The most revolutionized technology, which changed the whole digital currency space by introducing platforms such as bitcoin is blockchain. It is a distributed ledger which has the capability of maintaining a log of all the transactions occurring in the network. Recently, IoT is captivating a significant research interest. However, it can be observed that in areas such as building a distributed, secure and without third party environment for IoT, blockchain can be a beneficial tool. This paper aims to bulid a extensive view of different benchmark contributions in this direction. Initially the basics of blockchain and its application toward auditability, security, and characteristics of decentralization is discussed. Further, current challenges of centralized IoT models and their solution using blockchains for achieving decentalized and secure medium for the IoT are introduced.

1 Introduction

IoT was first introduced in 1999 by Ashton [1]. In 2015, a document was realized by the IEEE IoT with aim to establish a clear definition IoT with its applications in small to large and distributed complex systems. The document provides a clear picture of requirements of IoT architecture and technologies to enable it. It clarifies that IoT consists of number of hetro/homo devices which can sense and collect data which can be computed and analyzed for facilitating human uses in several differ-

S. Sharma (✉) · R. Chaudhry
Department of Computer Science and Engineering, SPM-International Institute of Information Technology Naya Raipur, Naya Raipur, India
e-mail: shivani@iiitnr.edu.in

R. Chaudhry
e-mail: rashmi@iiitnr.edu.in

D. Bhardwaj
Department of Electronics and Communication, Thapar Institute of Engineering and Technology, Patiala, India
e-mail: dinesh.bhardwaj@thapar.edu

A. Nanda and N. Chaurasia (eds.), *High Performance Vision Intelligence*, Studies in Computational Intelligence 913,
https://doi.org/10.1007/978-981-15-6844-2_3

ent fields. Iot is growing to a fast pace due to ubiquity of Internet, availability of wireless devices, cheap embedded computers, etc [2]. Large number of researchers have their significant interest in IoT which is causing reduction of size and improvement in smartness. Different domains are exploring the capability of these devices such as housing, precision agriculture, infrastructure, healthcare, vehicles, etc. This data generated by these smart devices need to be analyzed but may contain come private/confidential information which may lead to serious privacy threats. Further many of the existing infrastructures are centrally operated, hence, facing issue of single point failure. These issues hinder scalability and hence hesitation in users in adopting IoT widely. Further, centralized infrastructure leads to higher latency which cause hindrance during its application on smart grid and smart cities.

Today research toward empowring IoT helps in alleviating the issues for improving the privacy and security with in fog, mist, and central authority network architectures. De-centralized approach has become a new solution which allows the long-term IoT growth by preventing single point failure. Existing methods are for preserving privacy, data handling makes it essential for the high end clients which are escorted by a third party, hence they need to trust this third party and provide collected data to them. These third party are semihonest and may misuse the confidential data in anyway. The centralized network architecture faces the following challenges:

- The failure of centralized servers would make whole system to go down as entire infrastructure will be paralyzed [3]. Further, any attempt to denial of service (DOS) attack if become successful on such centralized servers would lead to entire system failure.
- Data shared and stored over a centralized server is always at the risk of exposure which could lead to number of privacy threats. The confidential information such as heath records, transaction records, habit, etc may get misused by un-trusted party. Data owners after sharing data to central server do not have much control over its accessibility.
- Traceability and accountability are another factors lacking in case of centralized cloud. Centralization requires a mandatory trusted third party for data storage, handling, and analysis which results into deletion of tampering of data.

Exponential growth in IoT made centralized servers to loose their efficiency while handling large number of end-to-end communications which is a necessary part for IoT automation function. Hence centarlized approach acts as a bottleneck in the growth of the IoT. These issues made it compulsory to rethink about the structure of IoT. Currently, 'Blockchain' is one of widely adopted technology for providing secure and distributed environment for IoT [4]. Blockchain was instigated by S. Haber and W.S. Stornetta of 1991. They quoted that "a cryptographically secured chain of blocks" [5]. However, got its reorganization in 2008 and 2009 when S. Nakamoto, an anonymous person formally defined it [6] and implemented it in crypto-currency called Bitcoin. It serve as a public legender for all network transactions [7].

Today many industries are using blockchain as their mainstream technology in various domains such as finance, logistics, agriculture, insurance, etc. This technique is a promising solution with the ability of transaction digitization. It is a paradigm shift in providing transparency in number of processes. If we observe the technique from higher perspective, blockchain uses cryptography to prove trusty environment in purely distributed environment, hence, nodes which are conducting data can reach rapid pacification. All the features of blockchain technique are foundation features of a serverless and record keeping approach, hence, number of researchers show their significant interest in devloping methods to levearge blockchins to decentralize the communication in IoT and remove the need of any centralized authority for maintaining the securiity prespective. The agglomoration of blockchain and IoT has emerged as a new interest as it possesses the following potential benefits:

- Main achievement is in terms of enhance fault tolerance and removal of singular failure point after shifting from centralized system to blockchain-based IoT. This also facilitates in removal of bottleneck issue faced during growing IoT reliant when working with centralized servers [8]. Furthermore, no thid party is required for the sake of handling confidential data of IoT users.
- IoT device autonomy and end-to-end device communication becomes easy in decentralized network, as the devices need not to pass through centralized server iteratively for performing any automation service. The integrity of data and identity of participants can be easily verified. software updates to IoT devices can be deployed securely due to tamper proof and protected storage in blockchain.
- Due to the absence of any central authority the contents, data, and event logs of blockchain retain immutability. This guarantees accountability and traceability. Further reliability and trustlessness are another important achievements attained via blockchain.
- Usage of smart contra for programmable logic is another important contribution of blockchain [9] through which IoT interactions can be treated as the transactions. This helps in enabling functions such as controlling the assessing authority, authentication, maintaining the confidentiality for the purpose of improving the privacy trait in blockchain-based IoT.
- Blockchain is a gateway for number of opportunities for service monetization. The trustlessness in the network based on blockchain helps in secure micro-transactions during IoT services and data.

2 Blockchain

Blockchain is basically the combination of best features of cryptography, public key, economic modeling which can be applied to peer to peer network. This helps in achieving synchronization in distributed database [10, 11]. Furthermore, blockchain is a form of data structure which posses the ability to store the transaction occurring in a network [7]. In applications such as cryptocurrencies where any form of data may get exchanged have high utility of blockchain. In peer to peer network all the participants maintain an identical image of the ledger. Decentralized consensus algorithms are used to add any new information of the transactions if arrive. In this

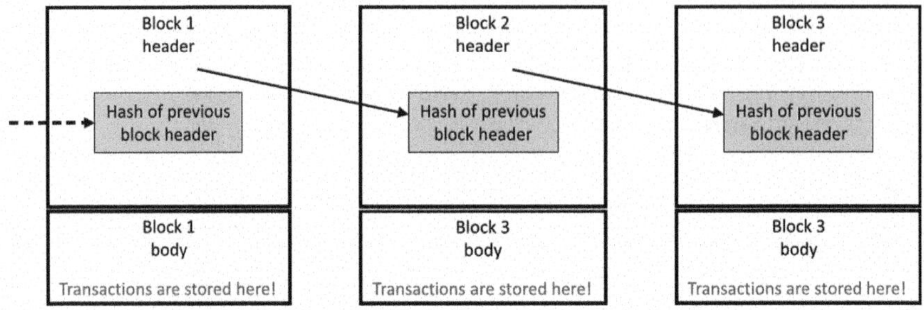

Fig. 1 Logical Blockchain Representation

section a discussion about the salient features, basic structure, and working principles is made.

2.1 *Salient Features of Blockchains*

- Decentralization
- Immutability
- Auditability
- Fault tolerance

2.2 *Blockchain Structure*

Several blocks combine to form a blockchain. Each block maintains details of the transactions which has occurred within the network. Transaction can be called to any transfer of data or token between the users in network. Each block can be logically viewed as combination of header and body. Each block in the body contains the transaction and header has the previous block identifier and the block's own identifier. Due to the existence of header, each block is coupled together in a link format just like the linked list. A logical view of blockchain is shown in Fig. 1. The very first block of the chain is called "genesis" block [12].

Each block is cryptographically hashed to get its identifier. This makes the blockchain contents immutable. Further, the header also contains timestamp of the block, i.e., time when the block was published as well as the Merkle tree root with respect to all the transaction which are cached with in the single block of the blockchain [13]. The Merkle tree root stored helps in significant reduction in efforts which are required with the purpose of validating a transaction have being in a particular block. Block chian is a linear data structure. Higher number of transaction makes the space for new blocks. Transaction within a block have their own

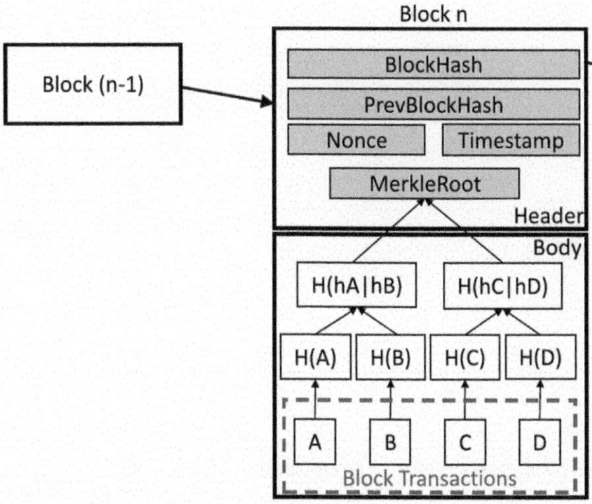

Fig. 2 Block header and Markle Tree for transaction storage

ID. These ID's are also obtained by cryptographically hashing the information of the corresponding transaction stored with in a block. These ID's of transactions are hashed in a pair and stored in a hashtree format in the block as shown in Fig. 2.

2.3 *Types of Blockchains*

1. Public Blockchains: These blockchains are decentralized in nature means all the participating members can add new blocks and can access contents of blockchain. These blockchains are also termed as permissionless as anyone can maintain a copy and can get engaged in validation process of new blocks.
2. Private Blockchains: Unlike public blockchains, these are permissioned means and every participating node is a well identified associates of a centralized organization. These blockchains are most suitable for single organization. In cases such as when different departments or individuals require to exchange data then private blockchains can be used to keep track of such exchanges in a synchronized distributed database. They do not require any extra overhead of currency or tokens to function. Furthermore, they do not require any processing fee.
3. Consortium Blockchains: They are also called as federated blockchains and are very similar to the private blockchains means they are also permissioned. However, unlike private they can span multiple number of organizations with maintained transparency among the participating parties. It can be used as synchronized, auditable, and reliable distributed database. All the data exchanges between the involved consortium parties are tracked.

	Public Blockchain	Private Blockchain	Consortium Blockchain
Participation in Consensus	All nodes	Single organization	Selected nodes in multiple organizations
Access	Public read/write	Can be restricted	Can be restricted
Identity	Pseudo-anonymous	Approved participants	Approved participants
Immutability	Yes	Partial	Partial
Transaction Processing Speed	Slow	Fast	Fast
Permissionless	Yes	No	No

Fig. 3 Comparative analysis of different types of blockchain [82]

A comparitive analysis of the different types of blockchain can be seen in Fig. 3.

2.4 Smart Contracts and Consensus Algorithms

The set of programmable applications which tends to manage the transactions under given set of specifications and conditions within a blockchain are called smart contracts. They are just equivalent to traditional economic contracts between involved parties. These costracts do not have any centralized authority or any intermediate party to check if the conditions in a smart contract are fulfilled or not.

Smart contracts are used perform different functions with in a blockchain such as

- With majority of participants agree to sign a single transaction called multi-signature transaction [14]. Such transactions are allowed by smart contracts.
- It enables the automated transactions which are triggered by some specific event.
- It provides utility to other existing smart contracts.
- It allows storage using membership record, lists, etc for keeping the information specific to the applications.

Consensus algorithms: These are the algorithms which help in securely updating the replicated shared states. They are essential for proper working of blockchain. Systems which are based on "state machine replication" within a blockchain are ensurely synchronized and in agreement by using consensus protocol. Consensus algorithms are a most captivating research topic nowadays.

3 IoT Privacy Using Blockchain

3.1 Privacy Concerns in Centralized IoT Models

Iot is a source to new capabilities when viewed from consumer level, for example, smart television, smart home, voice command to home computer to switch off the lights, commands for temperature maintenance, and next day groceries order by a smart fridge. These all examples work smartly by processing the data collected by different sensors. The collected data facilitate us to work smart but with the risk of privacy leakage as all these analysis is done by a centralized server. The privacy issues arise not only at the point of analysis but also at the time of storage, collection, and transmission too.

The collected data can be used for several diverse purposes such as an organization may collect usage data of their equipment which were given to others on a lease for the billing purpose. Data may be analyzed for finding the patterns of customers in retail store for deciding the discounts [23]. Customers take risk by providing their personal information to the organization when filling Internet-based forms and applications. These Internet-based thing do not provide any detail to the customer that for what purpose data they are providing will be used, stored, transmitted, or shared with third party [24]. In worst case it may be a mass-surveillance program [25]. While sharing data user must have to trust that their data will be transferred securely and will be used with all integrity and confidentiality maintained. Any unsecure data anlysis, storage, or transfer may lead to the loss of confidential information which can be misused in anyway [26]. Apart from security concerns during authentication and computing, it is a provocation to implement strategies which make it sure to maintain the data integrity, ownership, confidentiality, and governance [27]. In [27], "Privacy by design" has been discussed for empowerment of the users and making them able to control which data to be collected and stored. The goal of design is to implement access control policies for deciding weather to allow the access or not. In case of policy violation by any rogue sensor network, currently the solution is that the users which are going through the broker [28] can be subjected to threat as the broker is himself a intermediate entity between sensor and network. Furthermore, other existing techniques which are dependent on traditional centralized infrastructure of IoT such as signatures [29] and ring signatures [30] use centralized intermediaries which are then self-vulnerable to privacy threats. Both these techniques, i.e., group and ring signature follow the same concept. The data is transferred by the user via broker in order to hide the user identity. Another solution exists is k-anonymity [31] for preserving the privacy. The basic idea is to suppress the probability of any attribute to be detected by n times. However many research later shows that the idea of k-anonymity does not adequately assure the privacy of IoT data [32]. The main reason for the same is any common attribute in an anonymized data can be back tracked to find the personal information. For example, in a hospital data, intersected features of a specific disease can be tracked in the backward mode for evaluating the related medical informatic details of any specific patient matching to the particular common attribute. If we shift the paradigm toward decentralized technology we can observe that the number of reserach work have been done for using the blockchain technique as a promising solution to achieve "privacy by design".

3.2 IoT Privacy Using Blockchain-Based Decentralization

In last decade, decentralization have made its impact in sorting out the issues of privacy. Alcaide et al. [51] discussed the pre-blockchain solutions for achieving the authenticating of the anonymous parties in distributed environment. The solutions are based on "cryptographic Zero-Knowledge Proof of Knowledge (ZKPK)". However, number of drawbacks were counted for this solution such as prone to attack when actual user is impersonated by the adversary [52]. In recent time, blockchain is captivating the research's attention for decentralizing the IoT. It lays the foundation of decentralized network and secure data transfer without necessitating the need of authorization and authentication in intermediate stages. Its immutability feature over records provides a reliable solution for privacy preserved data sharing and micro payments in IoT. Therefore, it can be stated that, the blockchain merged with privacy preserving network design of IoT is an most active as well as fertile area of research.

All blockchain-based interaction taking place are publicly available as well as verfiable. Therefore, the IoT data which is either on-chain or off-chain has to be in encrypted formatas well as the access policies are also applied on blockchain. The very first step for developing a "private by design" solution is ensuring the user's data ownership such that the user can control the flow of his data means when and by whom to access. Users are also facilitated to decide if they want their data to be privated and kept in decentralized storage medium. For data ownership Zhang and Wen [33] presented a token access method. The method is based on the idea that the people can issue a transaction to the users for access of their own encrypted data. In such cases, the user can have a full control in deciding what data they really require to swap in place of the services or monitary incitements. This helps them in controlling their IoT data. Furthermore, [34, 35] provides solution for controlling the private ownership of the IoT data. The idea is that the data owner should have the complete authorization to decide which party should be granted to access authority of their IoT devices. FairAccess uses the concept of smart contract. The IoT user can discriminate between the parties for equating the role-base rights to the parties appealing for the data access. This is done on exchange to the service or monetary incentives. Similar tokenized techniques have been presented for access grant to the requesting candidates using the IoT data owner's discretion, in parallel to this the data in such methods are stored off-line using Decentralized Hash Tables (DHT).

Another frame work called PISCES framework is presented in [37] with the goal to provide "privacy by design". The aim have been achieved by enforcing data governance and ownership. The Privacy Validation Chain (PVC), roles of data controller, and data provider have been defined in order to maintain audit able track of data usage events. Additional PVC blockchain allows IoT users to have proper right over their data. A blockchain-based platform is proposed in [38] called as PlaTIBART. The platform is meant for the applications which are based on data interaction. It is source of various tools and techniques for managing and deploying IoT blockchain applications in private blockchain. The private blockchains are used because of its privacy feature and ability of fast transaction as well as off-chain communication implementation for private data transfer. Another solution for offline data storage and sharing is presented in [39]. The authors propose to use private blockchain for logging hashes of data chunks which are stored in a storage platform. This is based on Trusted Execution Environment (TEE). Further, they considered Intel SGX as a trusted environment for ensuring privacy of IoT data and application code. Cha et al. [22] presented a method to manage the IoT device and the issues of data transaction over blockchain by using the blockchain connected gateway. The gateway maintains privay awareness. The blockchain stores records in immutable encrypted format and enhances the privacy on IoT edge by using several BLE devices.

Another method proposed in [41], introduces software defined cloud computing. The method used blockchain-based access control for providing distributed privacy solution. The method presented in [42] is with the aim of providing privacy preserving access model. The blockchains and fine grained access control policies are presented which help users to govern their data. In [43], Rahulamathavan et al. propose to use a attribute-based encryption technique for encrypting the sensor data and enabling the privacy in IoT-Cloud environment. Chen et al. [44] came up with a jointCloud approach. It is a hybrid method called cloud-blockchain for achieving privacy for the IoT data. They are basically private cloud used for storage of data and a overlay blockchain for collecting the data transfer and IoT interaction event records. The Joint Cloud Collaboration Enviornment (JCCE) serves as a agglomerative way between private clouds. It consists of the JointCloud Blockchain (JCB) which helps in managing the transactions, supervision as well as community functions using smart contracts. The private blockchain helps in forming collaborative medium over cloud storage with the additional cost of server maintenance. It helps in maintaining tiered architecture where private blockchain gets connected to public blockchain. The owner of private blockchain can decide over communication with selective requester or other private blockchains. Hardjono and Smith [45] gave a privacy preservation method for commissioning IoT devices over a cloud. The idea is to use permissioned blockchains. The idea is to grant set of resource constrained IoT device without the exposing the identity. The solution is based on the ChainAnchor system [53] which aims to provide pseudonymity with in permissioned blockchain. This uses ZKP scheme and enhance privacy IO (EPID) [54].

It can be stated that the most promising solution for achieving the the "private by design" IoT is to use the tired architecture for blockchains. Either by using multiple private blockchains connected to a public blockchains OR inoperable blockchains which are connected to the other blockchains for forming a network. The users in different blockchains can slectively choose to express data with another blockchains. Dorri et al. [86] proposed a new architecture for privacy preserving in smart home where owners can log IoT events. The user can decide to share any data amount out of their encrypted data with others using a public overlay blockchain in accordance to the access control policies present in block header.

Smart contracts have been introduced by Hawk [15] which can help in implementing the programmatic access-control mechanism using smart contracts. Author accounts for sensitive and non-sensitive information by providing the degree of privacy. Conoscenti et al. [40] proposed to use peer-topper storage to remove the issue within blockchain.

The privacy of IoT user data can be enhanced by using peer to peer storage for off-chain data. Blockchain supports to store encrypted hash of the transnational information and data. Ali et al. [46], presented a multi-layered blockchain architecure, which is based on the concept of smart contract-based access and peer to peer storage. The solution is based on IPFS for storing data in distributed medium. Boom in renewable energy solutions make power grids to experience wide changes. Decentralized applications of the IoT which are helping in managing the transitive microgrids are emerging. Much work is being done in terms of energy saving applications which require privacy, monetigation, and decentralized control. Aitzhan and Svetinovic [16], proposed the concept of group signature and off-chain encryption of the anonymous messages in order to provide privacy in application such as energy trading. Laszka et al. [17], in recent times proposed a solution for preserving the privacy of personal information of energy producers while enabling them tokenize and trade units of energy with consumer. The new public private key pair helps producer to achieve anonymity in total in terms of transactions which are generated as well as maintained by a private broker. The smart contracts have been used by the authors in [18, 47] for achieving privacy. The tariffs for sharing the energy with in the smart grid is decided for getting a cost-effective way. Wang et al. [48], presented an incentive way for preserving the privacy in crowd sensing applications.The concept of k-anonymity is used to make sensing stream private. The authors use the node cooperation verification for making k-anonymity possible. The K nodes form a group which in cooperates for meeting the requirement of k-anonymity. Although the idea of k-anonymity have several disadvantages the proposed approach manages to resist impersonation attacks but no explanation is provided in terms of collusion attack. Recently, pseudonym management solutions have been used widely for IoT applications for enabling privacy in blockchain. However, fixed singular pseudonymous addressing does not provide enough of privacy even in the situations where off-chain data has to be transferred in a network. Singular blockchain can also be back-traced publicly which can cause to reveal the personal information of user such as id's [55]. In recent times a number of algorithms have been proposed for changing and updating the pseudonymous address [56]. Researchers shown significant interest in developing pseudonymous address management using fog computing for preserving connect vehicle privacy [57]. The identity of a singular vehicle existing with in a network have been masked by using a wide range of pseudonymous. This have been proved a significant boost in privacy [58]. Pseudonymous address management has been widely exploited for endsuring the transactional level privacy without any third party intermediaries. Kang et al. [49], focused on the solution to preserve the privacy for data sharing in vehicular networks while maintaining all privacy features. The updating mechanism of pseudonymous address helps in preventing a single vehicle which is being tied to a single blockchain address. Kang et al. [19] and Li et al.[20] pointed a new approach for connected hybrid vehicles in consortium blockchain by using pseudonymous address updating. They propose to implement an improved version of proof-of-work consensus mechanism but with less constraints. The local aggregators perform block validation and can be accountable if in case any wrong block creation is detected. Lu et al. [59] present use of pseudonymous address updating for acheing the privacy of VANETs. They also propose to maintain authorization and record messaging in every different blockchains for the purpose of adding auditability. Gao et al. [50] presented a new approach by using Hyperledger blockchain for implementing the secure payment mechanism particularly for vehicle-to-grid network. The use of Hyperledger PBFT does have the limitation of scalability of a network, however, it can acheive higher efficiency and speed of transaction. Different registration records are managed which could be visible to only authorized users for auditability.

3.3 Privacy Solutions in Industry for Secure IoT

Lola Cloud,10,a home intelligence system opted for an interesting solution to tackle the issue of privacy. They kept their users accounts and storage private by using blockchain smart contracts. COSMOS11 is another blockchain project with the aim to connect blockchains horizontally such that all contents of blockchain remain private even during interaction with other blockchain. Supply chain solutions are also very prominent to maintain immutable private records. For example, Provenance12, which relies on auditability of blockchain to ensure traceability and transparency of the food market products. This sector of supply chain is in advantage by using smart embedded devices which enables to push data autonomously in to the blockchain infrastructure. Hence, it create a tamper-proof and decentralized records in case of Skuchain13 and BriefTrace.14.

3.4 Summary and Insights

From the sections discussed so far that the blockchain helps in achieving auditability while providing all contents publicly accessible is a challenge in terms of privacy. We discussed number of recent contributions for warranting security in blockchain-based frameworks of IoT. The solution ranges from approaches which leverages smart contracts in possessing access policies to more advanced methods such as tiered blockchain architectures for energy transacting networks.

4 Trustless Architechtures for IoT

4.1 Issues of Trust in Centralized IoT Architectures

Cloud computing is a paradigm around which IoT services revolves. The IoT devices collect data which is processed and get stored in cloud only. However, the ubiquity of intelligent devices affects the IoT ecosystem. This is because the devices interacts with the users and collects the data which may be confidential to a user. Although, cloud computing provide data widely and make it accessible to others in real time and further this data is stored by a central entity. This storage and accessibility to a wide range of clients may cause misuse of the data. Recently, the researchers are moving into two different directions, i.e., one to improve the strength of trustful architecture by implementing more secure privacy preserving algorithms which could be used for dissemination and storage of IoT data. Second direction in which researcher are paying attention is toward "trustless" architecture [71]. These architectures depend on peer to peer approaches to validate the transactions occurring among participants. The first direction aims to include encryption for enhancing the trust factor in centralized techniques. This is complex for constrained IoT devices as it can make it necessary to either go for simple encryption techniques or not to use encryption at all for the purpose of communication. More secure encrption approaches such as AES-256 are definitely secure but induce the latency which definitely defer the application from the requirement of real-time solution. However, decentralized schemes remove the requirement of trusted third party and the peer to peer configuration positively supports the IoT environment by bearing the large availability of nodes with in the system.

4.2 Trustless IoT Architectures with Blockchains

Trustless architecture and synchronization combine to form blockchain. They maintain a chain of immutable transactions which is shared among different participates as discussed in previous sections, hence presents a trustable solution for the problem of centralization occurs in cloud computing. A study has been done by Sousa et al. [72] regarding secure Mutli-Party Computation (MPC) in blockchain with the aim of generating a trustless environment hyper-localized edge computations in the IoT fog. Ali et al. provided a comprehensive study focusing on how blockchain ensures that the entities involved perform the computation on integrated data collected by different participants without forming a trust relation with central authority.

Enigma, a peer to peer network [36], used blockchain to agglomerate number of users of blockchain for the purpose of storage and analysis without compromising the privacy of the data. Enigma also combines the different services of the permissionless blockchain in order to do some public tasks while performing several private computations in parallel. Liu et al. [60] presented a data integrity, decentralized verification framework. It is based on blockchain and uses smart contracts. This framework make Data Owner Applications (DOAs) and Data Consumer Applications (DCAs) for integrity verification of the data stored in cloud, which is a trustless environment. The another solution [61], for improving the trust level in transaction involved in blockchain is presented. This is done by using javacard secure elements. Inspite of 32 byte secret keys, cryptocurrency smart card has been used. This smart card is designed using JC3.04 standard platform. The realization of trustless verification of the transactions combined with blockchain in industrial point of view is presented in [62]. The proposed technique makes tasks to get computed on different IoT nodes as a distributed blockchain application. This helps in logging and storing of different transactions performed by the involved devices. This also helps in maintaining and diagnosing the number of issues which may occur on the computing nodes themselves. Furthermore, in [63], Boudguiga et al. propose a new mechnism which is decentralized and help in pushing updates on IoT devices. This uses blockchain for software updates onto the devices for protecting any malicious updates. This mechanism is free from the need of the trusted broker for the updates. All the updates propagated to the IoT devices through the blockchain guarantees the data integrity. In [64], another approach is presented by Di Pietro et al. The framework is decentralized which uses credit-based blockchain. The approach renamed it as obligation chain and it has built in reputation system. For the sake of reducing the delay within the transaction happening in traditional blockchains, the devices perform on credit transactions, and they need to pay the credit back which gets added to their reputation. This obligation chain is a step toward achieving scalability with in blockchain transactions while establishing trust between end-to-end IoT devices.

In [65] a trustless IoT architecture called IoTChain has been proposed. In this all IoT devices register themselves on a blockchain. This registration helps in securely data storage, organization, and data sharing without any trusted third party. However, the scenarios where frequency of transactions is too high, the scalability of blockchain has not been sufficiently addressed by the authors. The trustlessness has been demonstrated in end-to-end communication of the IoT devices. In the same direction, Psaras et al. [66], an edge centric solution has been proposed. This helps to establish an architecture which is trustless for the IoT. The gateways and devices issue the transaction to the blockchain during the same phase when the edge devices communicate in trustless manner.

Another solution called Trustchain have been introduced by Otte et al. [67]. It is a scalable solution with sybil resistance. It is for trustless IoT which replaces PoW consensus. This is done by an alternate method for finding trustworthiness of current peers in the network called NetFlow. The TrustChain uses parallel chains which are used to record transactions corresponding to each participant. It is used to find out if the existing peers are contributing actively for the maintenance of data integrity. It is also beneficial in determining the faults whenever the transactions in corresponding chains do not match each other. In such case, the TrustChain breaks its service to the responsible peer due to which all discrepancy has occurred. It can be stated that blockchain has been used in keeping transparency in records for cases such as supply chain. Recently, number of research contribution toward the traceable record keeping has been marked such as in [68]. They presented an architecture for traceable record keeping in context to the food supply chains. The BigchainDB [73] has been used which is a scalable and distributed DB with records which are publicly available but required to be protected. In [69] Ethereum smart contracts are suggested to be used for transparent as well as trustless record keeping in terms of pharmaceutical records. In use cases such as supply chain the trustless environment is particularly beneficial because related data always carry real values related to business which is surely confidential. If such data is exposed, it can lead to number of privacy related threats.

Another concern which is captivating the researchers attention is the suitability of IoT devices in blockchain. The idea led to the lightweight solution with respect to the IoT devices as well as blockchain nodes. A step toward such a lightweight solution is presented by Dorri et al. [21] by presenting a "scalable multi-tier blockchain model". This framework work in direction of IoT initiates distributed mechanism with in the participating parties which control the blockchains and additionally distributed throughput in terms of trust. This management helps in ensuring that the throughput allocated to the nodes is in coalition with total obtainable throughput. Another approach uses a fog layer over the devices as blockchain is proposed by Samaniego and Deters [70]. The use of constrained IoT end devices is replaced as the proposed architecture has the goal to enable the fog layer with the trust.

4.3 Blockchains for Enabling IoT Trust in Industry

All new industrial stratups are exploiting the advantages of blockchain for business model such as Xage Security,15 a distributed decentral method for presuming a trust in between all the nodes connected to the industrial network. It aims of making the industrial control decentralized instead of having a single trusted third party. Ubirch GmbH16 is another medium providing the notrary services in terms of IoT devices and the collected data. It provides trust among them. A private blockchain protocol called Multichain17 is another service which provides decentralized control on access to the devices which are registered on a particular blockchain. This protocol uses the concept of round-robin for running over decentralized consensus in order to get approval of transactions.

5 Blockchain-Based IoT Security

The IoT devices are growing exponentially and estimated to grow by 29 billion till 2022 [74]. Ubiquity of Internet increases the space for attacks and privacy threats resulting into improved and complex security measures [75]. The prime aim of IoT to achieve automation while preserving privacy against possible threats and attacks. In this section, discussion regarding several menaces encounter by central party-based IoT and their solution via decentralization is done.

5.1 Security Issues in IoT Models

Initial exigent in terms of security roots from enlarging edge of IoT. Edge sensoring nodes are failure points means at these nodes several attacks such as "Distributed Denial-of-Service (DDoS)" can be easily targeted [76]. Formation of a dishonest group of nodes can really leads to IoT service collapse as recently noted in botnet attacks [77]. Mirai botnet mounted a crucial attack by compromising several Iot devices and got successful in generating malicious traffic in terabytes per second [78]. Afterwards when the source code of the attack got public, number of other attacks followed mainly notified in October'16 which affected conventional subsites for number of hours.

Further, another threat in term of availability of IoT service gets generated due to heave centralized configuration [79]. Single point of failure affects availability, confidentiality, and authorization [80]. It fails in guaranteeing the data misuse or tampering. This leads to attack by identity spoofing, traffic, and route analysis on the confidentiality. In today's data driven economy no data misuse policy is crucially important to prevent misappropriation of data. These confidentiality attacks leads to integrity attck such as modification attacks and Byzantine routing information attacks [80]. In centralized IoT, data integrity is another concern due to challenges like injection attack. This is due to the fact that in the cases where decisions are made on the basis of in-stream data. Cases like alteration of data, theft, downtime can be exponential loss. Security is an important pillar when smart devices are required to interact and get involved in monetary transactions. Current centralized solutions must be agglomerated with blockchains for imposing the security policies, making records publically auditable, removal of dependency from any third party.

5.2 Blockchains for Providing IoT Security

Decentralization via public-key, desigining fault-tolerant infrastructure, protection against attacks like DDOS, auditability, and now security to transactive networks such as Bitcoin, everywhere blockchain is becoming a source to a appropriate solution. False authentication can be easily handled by blockchain-based IoT solution as any device in the network is binded to a particular address hence every transaction and its source can be easily verified. Protocol used in public blockchains help in obstracting attacks like DOS by demending fees for every empty transaction it incurs. Thus blockchain promises a better security mechanism and improved security services to the IoT. Blockchain has been incorporated in several fields for creating better privacy preservation solution with respect to each such as

- Providing Access Control
- Maintaining Data Integrity
- Ensuring Confidentiality
- Improving IoT Availability

5.3 Industry Solutions for IoT Security

Number of startups are addressing blockchain as a new security solution like SmartAxiom,18 which is an edge-oriented software concentrated on this technology, i.e., blockchain. This help in device identification, maintaining data integrity, data privacy as well as device authentication. In fields such as logistics and supply chain blockchain-based BlockVerify19 provide a protected solution [81]. It focuses on providing anti-counterfeit measures to the users. Filament,20 recently developed "blocklet" chip which helps in connecting industrial devices into a blockchain-based network.

6 Future Research Directions

This section highlights the important research directions in which new opportunities can captivate the research interest and help the society to grow positively. The main research areas are as follows (Fig. 4):

- Privacy in Permissionless Blockchains: Permissionless blockchains make the data transparent to all means public for the purpose of auditability. It has pseudonymous addressing of blockchain. Privacy is preserved by providing total anonymity.
- Scalability in Blockchains: The transaction throughtput is reduced due to decentralized consensus in permisionless blockchain. Further, it requires large storage capability due to network-wide transactions.
- IoT Edge-Device Constraints: Resource-constrained IoT device needs to directly issue the transactions to the blockchain which in-turn require more computation complexity. Furthermore, all the IoT gateways connected to blockchain need computation and storage in abundant to sustain in blockchain network.
- Trade-Off in Public–Private Blockchains: Nodes in Public blockchain grow in thousands and provide guaranteed auditability and immutability. However, reduction in throughput and spike increase in latency require some research improvement. On the other hande the private blockchain produces high throughput but lacks in terms of accountability. Further the number of nodes can grow up to tens.
- Security Standards for Scripting Smart Contracts: Some official standards are required by the blockchain-based IoT for the sake of generating secure smart contracts but without exploiting them for any critical cause.

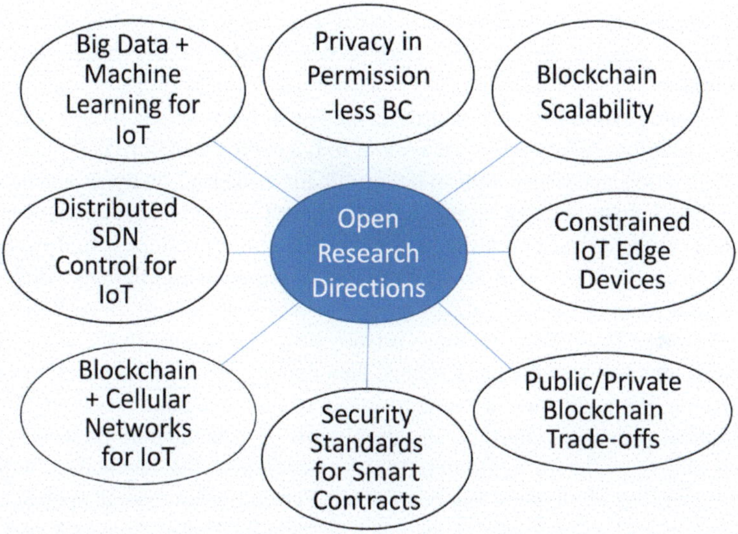

Fig. 4 Research Fields For Agglomerating Blockchains With IoT [82]

- Blockchains and Cellular Networks for the IoT: When cellular network is decentralized at the IoT edge will definitely affects the security of cellular network as well as at application layer of blockchain.
- SDN Integration for Blockchain-Based IoT Edge: SDN is more sensitive toward several privacy threats as it have typically heavily centralized control planes. Blockchain can be an effective solution to decentralize and secure these SDN planes.
- ML and Big Data for Distributed IoT Frameworks: Number of big data applications are required to be safe and secure making the room for blockchain technique for the same. Users are required to share their data on an open-source repositories, for the purpose of data analysis to make better machine learning models and IoT automation. Hence this shared data must be as secure as possible.

7 Conclusion

In this chapter, a deep study of the crucial and latest contributions in research have been brought up in the field of blockchain technology to a level of ripeness, with a main sate of indistinctness on developing platforms based on blockchain, designing applications and providing services which are suitable for the new and emerging era of the IoT is presented. Initially we focused on the blockchain by discussing its key features which is a series of several verifiable transactions holding the property of immutability. Distributed consensus algorithms have been used to achieve the immutation and security within the records. This helps in providing the trustless ecosystem for keeping the records where no requirement of any trusted third party is necessary.

The decentralization in cryptocurrency network, made blockchain a potentially beneficial solution. In current framework of IoT, a centralized third party is required for authentication, accessing, authorization, etc. for managing, analysis, and handling the data. Blockchain helps in laying down the base for decentralization where there is no need of managing, assigning, and authorizing intermediaries. In this chapter, different research scopes with in the framework have been highlighted which could be used as a potential field of development in terms of decentralized blockchain-based solutions. These fields includes privacy preservation, trustless, and protected communication between the participants, data management, IoT data, and resource monetization. We conducted an in-depth study of the blockchain, its main features, and technical principles. Further, the discussion on several important and recent research efforts made for agglomerating the benefits of the blockchain with IoT to resolve various challenges has been done. Further, a number of open research directions are discussed which can be taken as open R&D areas in near future.

References

1. K. Ashton, That Internet of Things thing. RFID J. (2009)
2. A. Al-Fuqaha, M. Guizani, M. Mohammadi, M. Aledhari, M. Ayyash, Internet of things: a survey on enabling technologies, protocols, and applications. 4th Quart. IEEE Commun. Surveys Tuts. **17**(4), 2347–2376 (2015)
3. U. Kumar, S. Sanyal, Survey of security and privacy issues of Internet of Things. Int. J. Adv. Netw. Appl. **6**(4), 2372–2378 (2015)
4. S. Huckle, R. Bhattacharya, M. White, N. Beloff, Internet of Things, blockchain and shared economy applications. Proc. Comput. Sci. **98**, 461–466 (2016)
5. S. Haber, W.S. Stornetta, How to time-stamp a digital document. J. Cryptol. **3**(2), 99–111 (1991). Jan
6. S. Nakamoto, Bitcoin: A Peer-to-Peer Electronic Cash System (2018). http://bitcoin.org/bitcoin.pdf. Accessed 12 Dec. 2018
7. F. Tschorsch, B. Scheuermann, Bitcoin and beyond: A technical survey on decentralized digital currencies. 3rd Quart. IEEE Commun. Surveys Tuts. **18**(3), 2084–2123 (2016)
8. P. Brody, V. Pureswaran, S. Panikkar, S. Nair, *Empowering the edge practical insights on a decentralized Internet of Things* (IBM Inst. Bus, White Paper, Armonk, NY, USA, 2015)
9. N. Szabo, Formalizing and securing relationships on public networks, First Monday, vol. 2, no. 9 (1997)
10. M. Pilkington, *Blockchain technology: Principles and applications, in Research Handbook on Digital Transformations* (Edward Elgar Publishing Incorporated, Cheltenham, U.K., 2015)
11. R. Beck, J.S. Czepluch, N. Lollike, S. Malone, Blockchain-the gateway to trust-free cryptographic transactions, in Proceedings of the 24th European Conference on Information Systems, (ECIS), Istanbul, Turkey, pp. 1–15 (2016)
12. P. Mueller, A. Rizk, R. Steinmetz, BlockChain a New Foundation for Building Trustworthy and Secure Distributed Applications (DAPP's) of the Future (2017). http://dspace.icsy.de:12000/dspace/handle/123456789/432. Accessed 12 Dec. 2018
13. R.C. Merkle, A digital signature based on a conventional encryption function, in Proceedings of the Advances in Cryptology (CRYPTO), pp. 369–378 (2000)

14. S. Omohundro, Cryptocurrencies, smart contracts, and artificial intelligence. AI Matters **1**(2), 19–21 (2014)
15. A. Kosba, A. Miller, E. Shi, Z. Wen, C. Papamanthou, Hawk: The blockchain model of cryptography and privacy-preserving smart contracts, in Proceedings of the IEEE Symposium on Security and Privacy, pp. 839–858 (2016)
16. N.Z. Aitzhan, D. Svetinovic, Security and privacy in decentralized energy trading through multi-signatures, blockchain and anonymous messaging streams. IEEE Trans. Depend. Secure Comput. **15**(5), 840–852 (2016)
17. A. Laszka, A. Dubey, M. Walker, D. Schmidt, Providing privacy, safety, and security in IoT-based transactive energy systems using distributed ledgers, in Proceedings of the International Conference on the Internet of Things (IoT), pp. 1–13 (2017)
18. F. Knirsch, A. Unterweger, G. Eibl, D. Engel, Privacy-preserving smart grid tariff decisions with blockchain-based smart contracts, in Sustainable Cloud, Energy Services (Springer, Cham, Switzerland, 2018), pp. 85–116
19. J. Kang et al., Enabling localized peer-to-peer electricity trading among plug-in hybrid electric vehicles using consortium blockchains. IEEE Trans. Ind. Informat. **13**(6), 3154–3164 (2017). Dec
20. Z. Li et al., Consortium blockchain for secure energy trading in industrial Internet of Things. IEEE Trans. Ind. Informat. **14**(8), 3690–3700 (2018). Aug
21. A. Dorri, S.S. Kanhere, R. Jurdak, and P. Gauravaram, Blockchain for IoT security and privacy: The case study of a smart home, in Proceedings of the IEEE International conference on Pervasive Computing and Communication Workshops (PerCom Workshops), pp. 618–623 (2017)
22. S.-C. Cha, J.-F. Chen, C. Su, K.-H. Yeh, A blockchain connected gateway for BLE-based devices in the Internet of Things. IEEE Access **6**, 24639–24649 (2018)
23. S. Sicari, A. Rizzardi, L.A. Grieco, A. Coen-Porisini, Security, privacy and trust in Internet of Things: The road ahead. Comput. Netw. **76**, 146–164 (2015). Jan
24. J.A. Stankovic, Research directions for the Internet of Things. IEEE Internet Things J. **1**(1), 3–9 (2014). Feb
25. G. Greenwald, E. MacAskill, NSA prism program taps in to user data of Apple, Google and others. Guardian **7**(6), 1–43 (2013)
26. C.M. Medaglia, A. Serbanati, An overview of privacy and security issues in the Internet of Things, in Proceedings of the Internet Things, pp. 389–395 (2010)
27. O. Vermesan, P. Friess, *Internet of Things: Converging Technologies for Smart Environments and Integrated Ecosystems* (River, Gistrup, Denmark, 2013)
28. G.V. Lioudakis et al., A proxy for privacy: The discreet box, in Proceedings of the International Conference on Computer Tool (EUROCON), pp. 966–973 (2007)
29. D. Chaum, E. Van Heyst, Group signatures, in Proceedings of the Workshop Theory and Applications of Cryptographic Techniques, pp. 257–265 (1991)
30. F. Li, Z. Zheng, C. Jin, Secure and efficient data transmission in the Internet of Things. Telecommun. Syst. **62**(1), 111–122 (2016)
31. L. Sweeney, K-anonymity: A model for protecting privacy. Int. J. Uncertainty Fuzziness Knowl. Based Syst. **10**(5), 557–570 (2002)
32. J. Domingo-Ferrer, V. Torra, A critique of k-anonymity and some of its enhancements, in Proceedings of the Third International Conference on Availability, Reliability and Security, pp. 990–993 (2008)
33. Y. Zhang, J. Wen, The IoT electric business model: Using blockchain technology for the Internet of Things. Peer-to-Peer Netw. Appl. **10**(4), 983–994 (2017)
34. A. Ouaddah, A.A. Elkalam, A.A. Ouahman, Towards a novel privacy-preserving access control model based on blockchain technology in IoT, in Proceedings of the Europe MENA Cooperation Advances in Information and Communication Technology, pp. 523–533 (2017)
35. A. Ouaddah, A.A. Elkalam, A.A. Ouahman, Fairaccess: A new blockchain-based access control framework for the Internet of Things. Security Commun. Netw. **9**(18), 5943–5964 (2016)

36. G. Zyskind, O. Nathan, A. Pentland, Enigma: Decentralized Computation Platform With Guaranteed Privacy (2015). https://enigma.co/enigma_full.pdf. Accessed 12 Dec. 2018
37. N. Foukia, D. Billard, E. Solana, PISCES: A framework for privacy by design in IoT, in Proceedings of the 14th Annual Conference on Privacy, Security and Trust (PST), pp. 706–713 (2016)
38. M.A. Walker, A. Dubey, A. Laszka, D.C. Schmidt, PlaTIBART: A platform for transactive IoT blockchain applications with repeatable testing, in Proceedings of the 4th Workshop Middleware Applications for the Internet Things, pp. 17–22 (2017)
39. G. Ayoade, V. Karande, L. Khan, K. Hamlen, Decentralized IoT data management using blockchain and trusted execution environment, in Proceedings of the IEEE International Conference on Information Reuse and Integration (IRI), pp. 15–22 (2018)
40. M. Conoscenti, A. Vetrò, J.C. De Martin, Peer to peer for privacy and decentralization in the Internet of Things, in Proceedings of the IEEE/ACM 39th International Conference on Software Engineering Companion (ICSE-C), pp. 288–290 (2017)
41. P.K. Sharma, M.-Y. Chen, J.H. Park, A software defined fog node based distributed blockchain cloud architecture for IoT. IEEE Access **6**, 115–124 (2018)
42. G. Zyskind, O. Nathan, A. Pentland, Decentralizing privacy: Using blockchain to protect personal data, in Proceedings of the IEEE Security Privacy Workshops (SPW), San Jose, CA, USA, pp. 180–184 (2015)
43. Y. Rahulamathavan, R.C.-W. Phan, S. Misra, M. Rajarajan, Privacy-preserving blockchain based IoT ecosystem using attributebased encryption, in Proceedings of the IEEE International Conference on Advanced Networks and Telecommunications Systems (ANTS), pp. 1–6 (2017)
44. W. Chen, M. Ma, Y. Ye, Z. Zheng, Y. Zhou, IoT service based on jointcloud blockchain: The case study of smart traveling, in Proceedings of the IEEE Symposium on Service-Oriented System Engineering (SOSE), pp. 216–221 (2018)
45. T. Hardjono, N. Smith, Cloud-based commissioning of constrained devices using permissioned blockchains, in Proceedings of the 2nd ACM International Workshop IoT Privacy Trust Security, pp. 29–36 (2016)
46. M.S. Ali, K. Dolui, F. Antonelli, IoT data privacy via blockchains and IPFS, in Proceedings of the 7th International Conference on Internet of Things, Art. no. 14 (2017)
47. F. Lombardi, L. Aniello, S. De Angelis, A. Margheri, V. Sassone, A blockchain-based infrastructure for reliable and cost-effective IoTaided smart grids, in Proceedings of the Living Internet Things Cybersecurity (IoT), pp. 1–6 (2018)
48. J. Wang et al., A blockchain based privacy-preserving incentive mechanism in crowdsensing applications. IEEE Access **6**, 17545–17556 (2018)
49. J. Kang et al., Blockchain for secure and efficient data sharing in vehicular edge computing and networks. IEEE Internet Things J. **6**, 4660–4670. https://doi.org/10.1109/JIOT.2018.2875542.
50. F. Gao et al., A blockchain-based privacy-preserving payment mechanism for vehicle-to-grid networks. IEEE Netw. **32**(6), 184–192 (2018)
51. A. Alcaide, E. Palomar, J. Montero-Castillo, A. Ribagorda, Anonymous authentication for privacy-preserving IoT targetdriven applications. Comput. Secur. **37**, 111–123 (2013). Sep
52. X.-J. Lin, L. Sun, H. Qu, Insecurity of an anonymous authentication for privacy-preserving IoT target-driven applications. Comput. Secur. **48**, 142–149 (2015). Feb
53. T. Hardjono, N. Smith, A.S. Pentland, Anonymous identities for permissioned blockchains, MIT Connection Sci. Rep. (2016)
54. E. Brickell, J. Li, Enhanced privacy ID: A direct anonymous attestation scheme with enhanced revocation capabilities, in Proceedings of the ACM Workshop Privacy in the electronic society, pp. 21–30 (2007)
55. A. Biryukov, D. Khovratovich, I. Pustogarov, Deanonymisation of clients in bitcoin P2P network, in Proceedings of the ACM SIGSAC Conference on Computer and Communications Security, pp. 15–29 (2014)
56. A. Boualouache, S. Moussaoui, Urban pseudonym changing strategy for location privacy in VANETs. Int. J. Ad Hoc Ubiquitous Comput. **24**(1–2), 49–64 (2016). https://doi.org/10.1504/IJAHUC.2017.080914

57. J. Kang, R. Yu, X. Huang, Y. Zhang, Privacy-preserved pseudonym scheme for fog computing supported Internet of Vehicles. IEEE Trans. Intell. Transp. Syst. **19**(8), 2627–2637 (2018). Aug
58. M. Raya, J.-P. Hubaux, Securing vehicular ad hoc networks. J. Comput. Security **15**(1), 39-68 (2007). http://dl.acm.org/citation.cfm?id=1370616.1370618
59. Z. Lu, W. Liu, Q. Wang, G. Qu, Z. Liu, A privacy-preserving trust model based on blockchain for VANETs. IEEE Access **6**, 45655–45664 (2018)
60. B. Liu, X. L. Yu, S. Chen, X. Xu, L. Zhu, Blockchain based data integrity service framework for IoT data, in Proceedings of the IEEE Internatrional Conference on Web Services (ICWS), Honolulu, HI, USA, pp. 468–475 (2017)
61. P. Urien, Towards secure elements for trusted transactions in blockchain and blochchain IoT (BIoT) Platforms, in Proceedings of the 4th International Conference on Mobile Secure Services (MobiSecServ), pp. 1–5 (2018)
62. A. Bahga, V.K. Madisetti, Blockchain platform for industrial Internet of Things. J. Softw. Eng. Appl. **9**(10), 533 (2016)
63. A. Boudguiga et al., Towards better availability and accountability for IoT updates by means of a blockchain, in Proceedings of the IEEE European Symposium on Security and Privacy (EuroS&PW), pp. 50–58 (2017)
64. R. Di Pietro, X. Salleras, M. Signorini, E.Waisbard, A blockchainbased trust system for the Internet of Things, in Proceedings of the 23rd ACM Symposium on Access Control Models and Technologies (SACMAT), pp. 77–83 (2018). https://doi.org/10.1145/3205977.3205993
65. B. Yu et al., IoTchain: Establishing trust in the Internet of Things ecosystem using blockchain. IEEE Cloud Comput. **5**(4), 12–23 (2018)
66. I. Psaras, Decentralised edge-computing and IoT through distributed trust, in Proceedings of the 16th Annual International Conference on Mobile Systems, Applications and Services (MobiSys), pp. 505–507 (2018). https://doi.org/10.1145/3210240.3226062
67. P. Otte, M. de Vos, J. Pouwelse, Trustchain: A Sybil-resistant scalable blockchain. Future Gener. Comput. Syst. **107**, 770–780 (2017)
68. F. Tian, A supply chain traceability system for food safety based on haccp, blockchain & Internet of Things, in Proceedings of the International Conference on Services Systems and Services Management, (ICSSSM), pp. 1–6 (2017)
69. T. Bocek, B.B. Rodrigues, T. Strasser, B. Stiller, Blockchains everywhere-A use-case of blockchains in the pharma supply-chain, in Proceedings of the IFIP/IEEE Symposium on Integrated Network Management (IM), pp. 772–777 (2017)
70. M. Samaniego, R. Deters, Blockchain as a service for IoT, in Proc. IEEE International Conference on Internet of Things (iThings) IEEE Green Computing and Communications (GreenCom) and IEEE Cyber, Physical and Social Computing (CPSCom) and IEEE Smart Data (SmartData), pp. 433–436 (2016)
71. S. Tai, Continuous, trustless, and fair: Changing priorities in services computing, in Proceedings of the European Conference on Service-Oriented and Cloud Computing, pp 205–210 (2016)
72. P.R. Sousa, L. Antunes, R. Martins, *The present and future of privacy-preserving computation in fog computing, in Fog Computing in the Internet of Things* (Springer, Cham, Switzerland, 2018), pp. 51–69
73. T. McConaghy et al., *BigchainDB: A scalable blockchain database* (BigChainDB, White Paper, Berlin, Germany, 2016)
74. W. Obile, *Ericsson Mobility Report* (Ericsson, Stockholm, Sweden, 2016). Nov
75. J. Granjal, E. Monteiro, J.S. Silva, Security for the Internet of Things: A survey of existing protocols and open research issues. 3rd Quart. IEEE Commun. Surveys Tuts. **17**(3), 1294–1312 (2015)
76. H. Suo, J. Wan, C. Zou, J. Liu, Security in the Internet of Things: A review. Proc. Int. Conf. Comput. Sci. Electron. Eng. (ICCSEE), **3**, 648–651 (2012)
77. C. Kolias, G. Kambourakis, A. Stavrou, J. Voas, DDoS in the IoT: Mirai and other Botnets. Computer **50**(7), 80–84 (2017)
78. E. Bertino, N. Islam, Botnets and Internet of Things security. Computer **50**(2), 76–79 (2017). Feb

79. J. Gubbi, R. Buyya, S. Marusic, M. Palaniswami, Internet of Things (IoT): A vision, architectural elements, and future directions. Future Gener. Comput. Syst. **29**(7), 1645–1660 (2013)
80. S. Sicari, A. Rizzardi, C. Cappiello, D. Miorandi, A. Coen-Porisini, *Toward data governance in the Internet of Things, in New Advances in the Internet of Things* (Springer, Cham, Switzerland, 2018), pp. 59–74
81. N. Kshetri, 1 Blockchain's roles in meeting key supply chain management objectives. Int. J. Inf. Manag. **39**, 80–89 (2018)
82. M.S. Ali, M. Vecchio, M. Pincheira, K. Dolui, F. Antonelli, M.H. Rehmani, Applications of Blockchains in the Internet of Things: A Comprehensive Survey. IEEE Commun. Surv. Tutor. **21**(2), 1676–1717, Secondquarter (2019)

Black-Hole and Wormhole Attack Using DYMO and AODV Protocol: A Review

N. Chaurasia, P. Dimri, and K. K. Gupta

Abstract From last two decades, mobile ad-hoc network has significantly contributed to the research. Mobile ad-hoc network basically filed of wireless ad-hoc network in which all the connected components shares their data and make communication with each other without any dedicated links/path or without wired communication. There are various protocols available for the secured communication and safely transfer the files and data. However, the attacker will always be in action to destroy the communication or link between the sender and receiver. In the following paper, we present the survey on various types of attack in the field of wireless ad-hoc network including internal and external attack.

1 Introduction

In the present scenario, wireless ad-hoc network is the most popular field for the researchers, wireless ad-hoc network is basically a field in which all participating devices or nodes communicate with other without and dedicated link or path, and in other words all the nodes are connected in a wireless modem or without wired. Wireless ad-hoc network is an area which contains some filed such as wireless sensor network, wireless mesh networks, and mobile ad-hoc network. Wireless ad-hoc network is a very popular especially for the applications like military services,

N. Chaurasia (✉)
Department of Computer Science and Engineering, Uttarakhand Technical University, Dehradun, Uttarakhand, India
e-mail: rgtu.nishant@gmail.com

P. Dimri
Department of Computer Science and Applications, G.B. Pant Engineering College, Pauri, India
e-mail: pdimri1@gmail.com

K. K. Gupta
Department of Information Technology, Rustamji Institute of Technology, Gwalior, India
e-mail: kamlesh_rjitbsf@yahoo.co.in

A. Nanda and N. Chaurasia (eds.), *High Performance Vision Intelligence*, Studies in Computational Intelligence 913,
https://doi.org/10.1007/978-981-15-6844-2_4

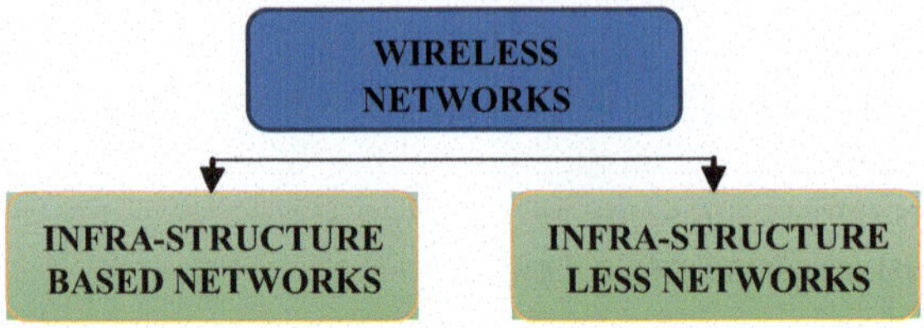

Fig. 1 Wireless network categories

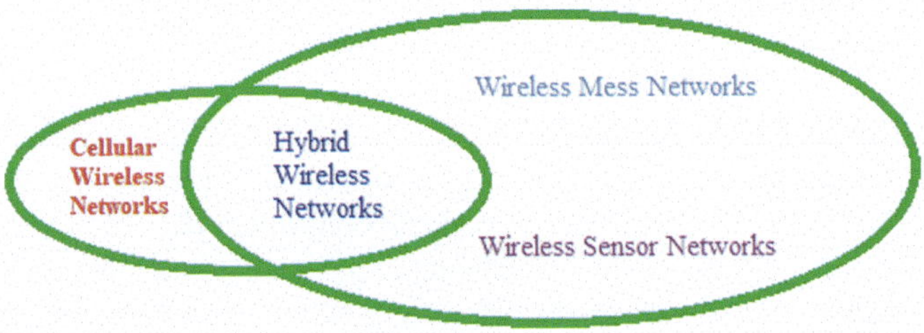

Fig. 2 Simplification of wireless network

automation control, intelligent traffic management system, security surveillance, etc. [2]. Wireless networks are basically classified into two categories as shown in the Fig. 1:

Infrastructure based Networks—these types of networks are having centralized system. In this network each and all devices are connected by the central access point. So each and every device is controlled by the centralized point. In this network, two types of devices are used, first works on fix infrastructure which provide control to another second type of devices that are worked on mobile position in the networks. These types of networks are opposite to infrastructure-based networks because there is no any centralized control on networks devices that no one centralized access point. Each and every device is connected peer to peer mode. All devices are self-configure, self-organized nodes, they can communicate with each other directly. Simplification of wireless network as shown in the Fig. 2.

Cellular wireless networks—this network is a replacement of high power transmission or receiving systems which is used on telephone systems. In this network, all nodes are connected without any physical connections or they are connected by wireless system. Examples of wireless networks are cellular mobiles phones, wireless networking system, wireless communication system, etc.

Wireless mesh networks—this network is a form of the wireless ad-hoc networks. These types of networks, all radio nodes are connected into the network by mesh topology. Many devices or nodes are connected to each other in the form of mesh format. So these networks are called wireless mesh networks.

Wireless sensor networks—these networks are the collection of small and cheapest nodes that nodes are filled with low memory, energy, and their process capacity. Wireless sensor networks hold large number of spatially dispersed, nice battery-

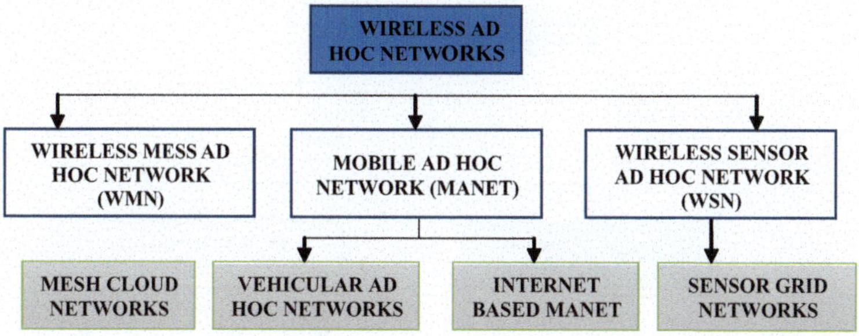

Fig. 3 Classification of wireless ad-hoc networks

operated embedded devices that produce process and transfer data to the users but these nodes have limitation and restricted computation capabilities and less processing. Wireless sensor networks have large number of sensor nodes, low data rate, and high redundancy compared to the MANETs.

Hybrid networks—hybrid network is basically combination of infrastructure-based network and infrastructure less network. Hybrid networks, each node is mobile but that nodes can be communicate may be infrastructure-based networks. Hybrid networks provide better network structure for coming next generations of wireless networks due to its QoS requirements of different technological applications [9]. In today's world Internet is a more popular term for sharing of information and data. In recent scenario, the more demand of computer increases the limit of transferred data form one station to another or from source to destination. The more demand of devices need more security function or tools for the network and the participating nodes or devices in a network, because there is always need to be kept secure and safe network from the attacker and other intruder. The classification of wireless Ad-Hoc network is shown in the Fig. 3.

Attacks in MANET can be classified based on its source, behavior, and nodes as shown in the Figs. 4 and 5. On the basis of source, there are two types of attacks that are external attacks and internal attacks [15]. While on the basis of behaviors there are passive attacks and active attacks. Routing protocol works with the Internet layer and they are defined as the act of transferring the data or information from one end to another end in a network. In the routing protocol various metrics decide the efficiency of the route in terms of number of hop counts, network traffic, and security of the data or information, etc. [6].

The main objective of routing protocols is to enhance the performance of network by measuring their performance parameters such as lessen end to end (E-E) delay, network throughput, improving the network lifetime (NLT), and maximizing energy efficiency. Selecting best path for routing and inter-network packet transfer are the two basic activities involved in routing [9].

Attacks in mobile ad-hoc networks are divided into two broad categories, one is passive attack and another one is active attack. In the passive attack classification, Eavesdropping attack and Traffic attacks are generally occurred and in the class of

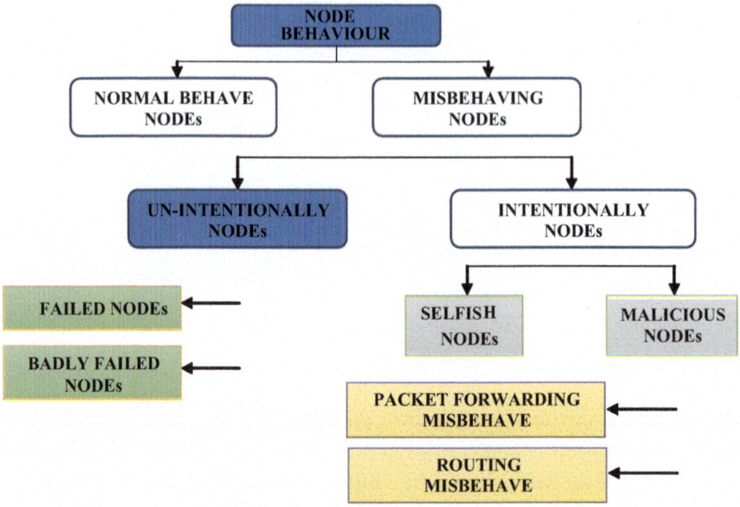

Fig. 4 Node behaviour in MANET

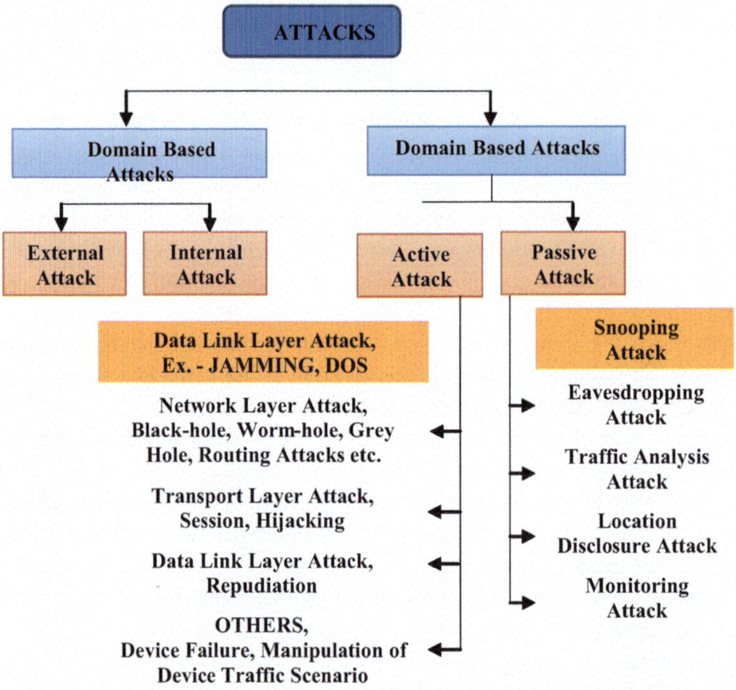

Fig. 5 Types of attacks

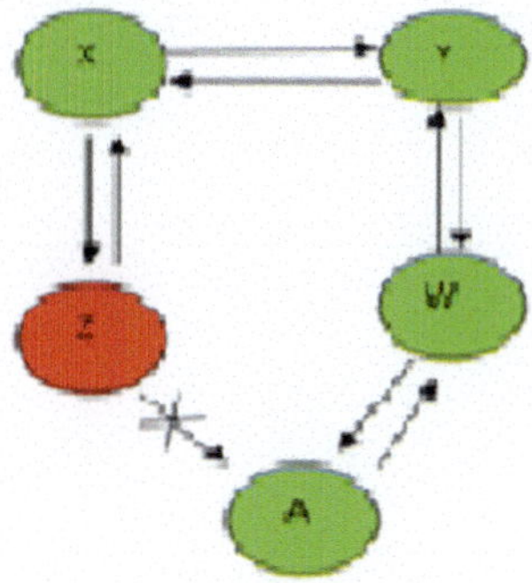

Fig. 6 Black-Hole attack [1]

active attack such as packet dropping attack, packet modification attack during the transfer of data packets, fabrication attack, and timing attack are directly related to sender and receiver timing at the time of packets sent or received [8].

Active attacks: Active attacks have an attack which exist in the network or basically transmitted data in the network. In this, an attacker change or alter the information packet that is being exchanged into the network. Network normal functionality affects due to active attacks. In the active attack, attacker or intruder can modify or update the data packets or drop the packet. It is very dangerous attack in the Mobile Ad-hoc Networks [10, 11].

Passive attacks: Passive attack is small stolen information attack because it is not affected on data packets or not done any alter or update types of operations on data. Due to its properties, it is very difficult to detect the passive attack in the network. Passive attack is only stolen or gather the information about the data types, network properties, communication properties between communicating parties. Some it may provide lead to active attack [7].

Black-hole attack: Besides the category of internal and external attack there are some types of attacks which actually occur in network during the transferring of data in a real-world environment. Black-hole attack is one of them among these types of attack generally occurred at the network layer in the model during the transferring of data from source node to destination node. Here, basically a host device sends the information or files from the one end, i.e., source end (here the hostile nodes are collected from all the packets) and send itself to another end, i.e., destination end. Here the hostile node or selfish node make conversation with all participating nodes and said to transfer the data securely in a minimum time using the short route [7, 11].

In the below figure, Fig. 6 shows the source node X and the destination node A. With some participating node Y and node W and the hostile or selfish node Z is depicted. When node X transfers the files for the node A, then node Z receives the file at early and send them to the node A, since node Z is hostile node so node A will receive the affected files or data.

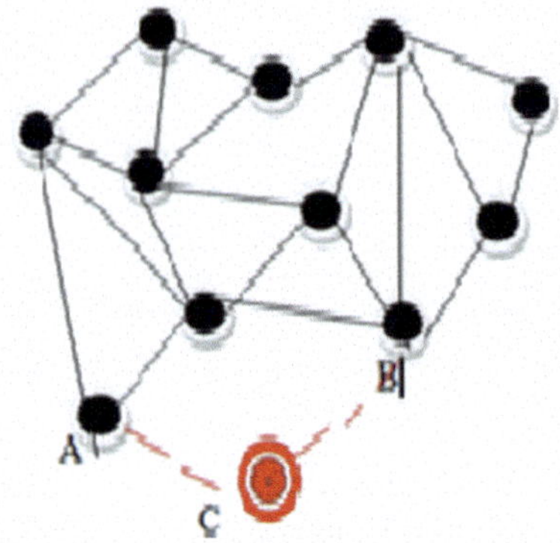

Fig. 7 Wormhole attack [8]

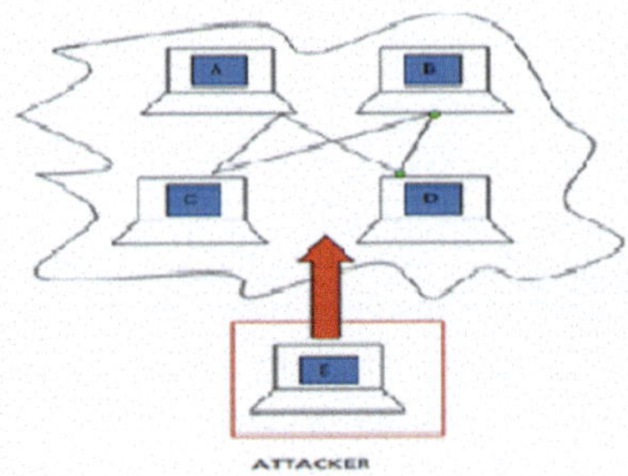

Fig. 8 External attack [8]

Wormhole attack: These types of attacks are generally occurred with the network layer when we are sending the data from one end to another end. In this attack category, generally it records the information or data from the sending end or broadcast to all participating nodes through the tunnel. This is supposed as the most dangerous attack for the network and very difficult to find and recover. Packet leases is a technique by these types of attacks can be detected [8]. In Fig. 7 below, a malicious node "C" is created as an extraneous link between the node A and node B. Here, the node "C" will be acting as a wormhole for the messages that are actually transferred between the node A and node B.

External Attack: There are various types of attackers present in the current scenario who are always ready to violate the whole system or network. External attacker is one of them in this category of attacker will attack in a network from the outside of the network as shown in the Fig. 8 [8]. Once they enter in a network, they will destroy the whole network. There are some tools for which we can prevent our network from such types of attacker such as firewall and intrusion detection or prevention system.

Internal Attack: In these types of attack category it is more difficult to predict the attacker and infected files or data in a network than external attack. In this attack category attacker will remain inside the network and affects and generates harmful files or programs for the network and destroy the whole network. These types of attacker (shown in Fig. 9) will play the role as an internal devices or nodes in a network and making a deal with every current node in network [8] (Table 1).

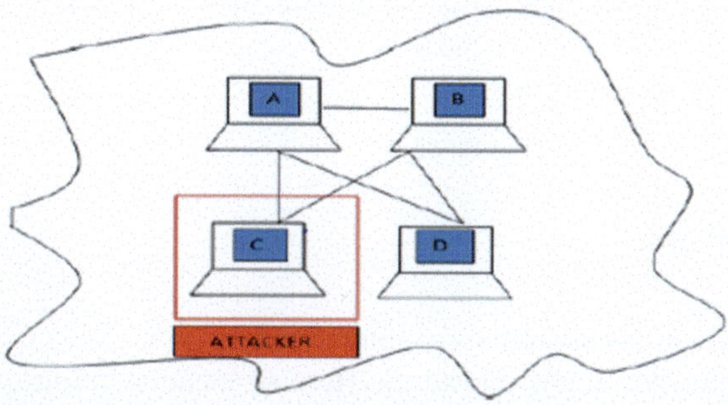

Fig. 9 Internal attack [8]

Table 1 Characteristics of MANETs

Characteristic	Description
Wireless	MANET network's nodes don't have any types of physical connectivity to each other
Heterogeneous	MANETs support heterogeneous types of networks that support scalable large network
High node mobility	Due to dynamic topology in MANET network provides high mobility in nodes
Shortage of storage	In MANET nodes work as self-organize and self-configure so they use tiny size of battery for energy
Autonomous behavior	In MANET, there is no type of centralized control, all the nodes work independently
Security, Routing, and Host configuration of distributed operation	Due to independent nodes all nodes participate in security, Routing in network, and configuration operations of network
Dynamic properties	MANET network works on dynamic topology to move freely and frequently in network
Natural and mobile behavior	Each node can move independently in MANET without any centralized control
Discrete connectivity	MANET nodes support discrete connectivity in the networks
Complete systematic environment	In MANET each node maintain proper procedure for all process in the network that all processes execute in systematic format way

2 Characteristics of MANETs:

Considering the study, following are the characteristics obtained in MANET.

The upcoming sections in the presented chapter includes introduction about the mobile ad-hoc network along with their attacks. Section 3 studies the rich literature for mobile ad-hoc network and the various routing protocol and possible attacks. Section 4, discusses about the comparative study for the routing protocol, attacks, using methods, etc. Section 5 is a briefing of challenges faced in the studied scenario and finally in Sect. 6, we conclude the about our paper which is based on the literature survey and specify the future scope.

3 Related Work

The following section is about the rich literature survey has been performed in the domain area with types of possible attack. Here we mention the survey followed by their reference number and their results. In paper [1], author introduced the detection as well as prevention strategy for black-hole attacks in environment. Here, author worked to avoid black-hole attack and introduce mitigation algorithm for the avoidance of malicious node and prevention of all the genuine nodes in the network. In Fig. 10, the authors depicted comparative graph for the normal node, black-hole attack detection and prevention of network form these attacks.

In the paper [3], the author introduces the comparison mechanism for the Dynamic MANET on demand and Ad-hoc on demand distance vector (DV) protocol for the congestion in a vehicular ad-hoc network. The DYMO protocol holds its ground from AODV protocol. In Fig. 11, the author shows the comparative analysis of throughput w.r.t. the AODV and DYMO protocol in the network.

This paper [4] is based on Cuckoo Search Optimization Algorithm, simulated for AODV & DYMO and analyzed the performance for Packet Delivery Ratio, End-to-

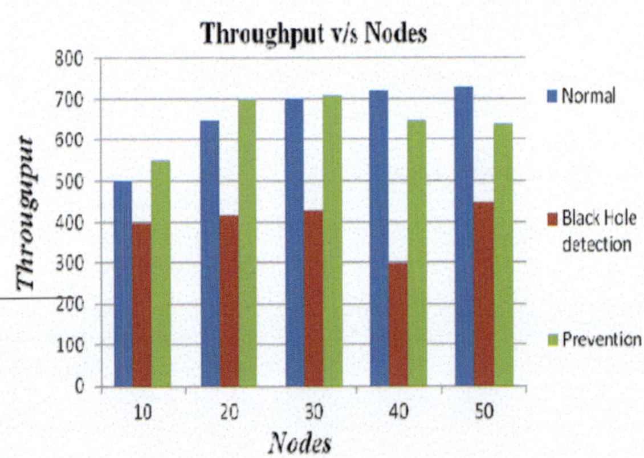

Fig. 10 Comparative result study for the throughput versus number of used nodes to detect normal, black-hole attack and prevention

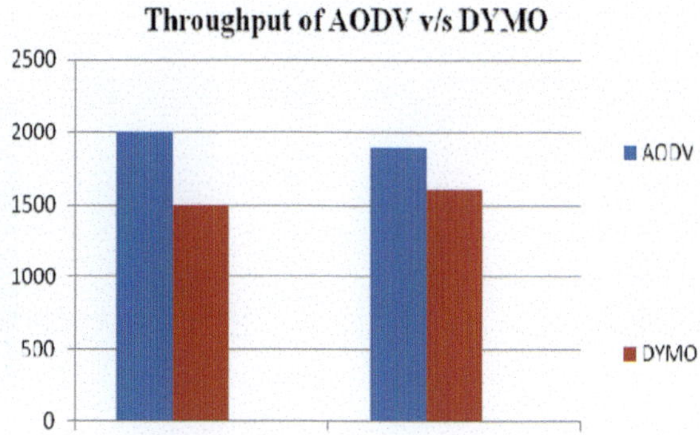

Fig. 11 Comparative result study for the throughput between DYMO and AODV protocol

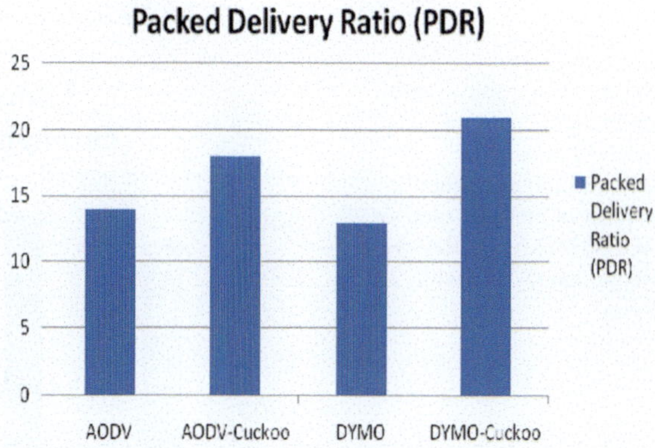

Fig. 12 Comparative result study for the performance parameter packet delivery ratio using the number of methods such as AODV, AODV-Cuckoo, DYMO, DYMO-Cuckoo

End Delay, and Energy Consumption using various simulation metrics. Better results were obtained for cuckoo search optimization algorithm based on AODV and DYMO in comparison to simple AODV & DYMO routing protocols, illustrated by Fig. 12.

The authors [5] present the performance comparison for the Dynamic mobile ad-hoc on demand protocol and Enhanced dynamic mobile ad-hoc network on demand protocol using the optimization techniques, viz, ant colony optimization. Here, the author selects the best short distance route with using ant routing technique. In the Fig. 13 below, the author showed the comparison between the performance parameter such as throughput measured in Kbps and the network size measured in the number of participating nodes for the Dynamic MANET, on demand protocol and Enhanced dynamic MANET on demand protocol.

In [7] this article author present the comparison between the Ad-hoc on demand DV protocol and Dynamic MANET formulated an on demand protocol in a 3-D (dimension) mode. Here author proposed a distributed algorithm for the specially Delaunay triangulation in an arbitrary sensor network. In Fig. 14, the authors showed

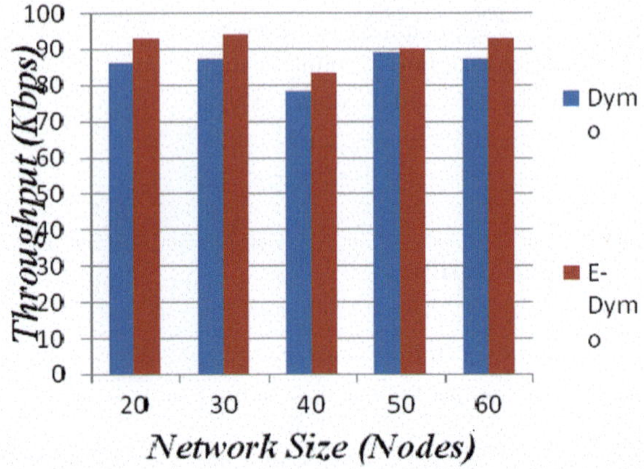

Fig. 13 Comparative result study for the throughput in Kbps versus network size (number of nodes) using the DYMO and E-DYMO Protocol

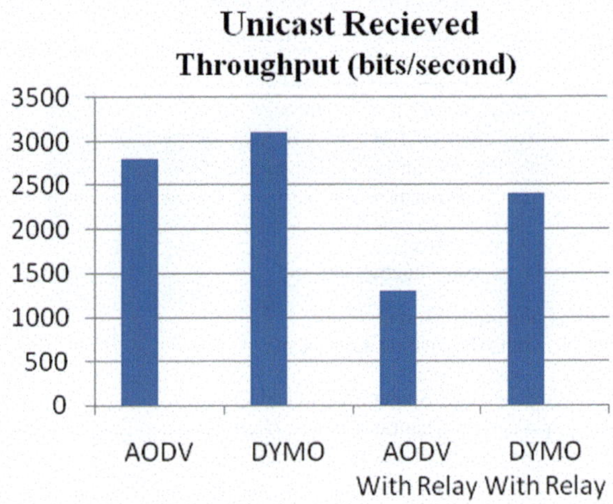

Fig. 14 Comparative result study for the uni-cast received throughput (bits/second) using the number of methods such as AODV, DYMO, AODV with Relay, and DYMO with Relay

the comparison between the performance parameter such as unicast throughput measured in bits/second for the Dynamic mobile ad-hoc network on demand protocol and ad-hoc on demand distance vector protocol with relay.

4 Comparative Study of Different Routing Protocol and Attacks

Following is the comparative study for the different routing protocols considered in this literature (Table 2).

Table 2 Comparative study of the protocols: DYMO and AODV with the various numbers of attacks

Ref. no.	Techniques used	Used protocol	Publication year
[1]	Author presents here the Detection and prevention strategy for Black-Hole attack in MANET	DYMO Protocol	2016
[2]	Here author presents the comparative performance assessment of Dynamic on demand MANET using the self-forwarding mechanism	DYMO Protocol	2017
[3]	Here author compare the performance of AODV along with DYMO protocol for message sent or received, throughput etc	AODV and DYMO Protocol	2017
[4]	Here author discussed Cuckoo Search Optimization based analysis of AODV and DYMO Algorithm using various simulation metrics	AODV and DYMO Protocol	2014
[5]	In this article author enhanced the performance of existing DYMO protocol using Ant colony optimization and increase the throughput also	AODV and DYMO Routing Protocol	2014
[6]	Here author presents the rich literature survey for the Dynamic on demand mobile ad-hoc network	DYMO Protocol	2014
[7]	In this article author discussed about the comparative performance analysis of ad-hoc on demand DV routing in addition to dynamic on demand mobile ad-hoc network routing protocol in the scenario of the distributed algorithm producing a Delaunay triangulation for an arbitrary sensor network	AODV and DYMO Routing Protocols 3-D Wireless Sensor Networks	2017
[8]	Here author proposed trusting AODV algorithm for identification (detection) and avoidance of wormhole attack as well as collaborative black-hole attack	Wormhole and Black-Hole attack using AODV protocol	2015
[9]	In this article author presents the comparative assessment survey for the various protocol used in a MANET, protocols compared with some performance parameter such as Protocol type, Routing Approaches etc	Compared the all MANET protocol	2016
[10]	The authors proposed a scheme for sending a control sequence to the neighbor nodes and expecting the nodes response	Gray and Black-Hole Attacks using Control Packets in MANETs	2017
[11]	In this paper author presents the intrusion detection on real time and also detect the wormhole attack using the Internet of things	Wormhole Attack Detection	2015

(continued)

Table 2 (continued)

Ref. no.	Techniques used	Used protocol	Publication year
[12]	Here author proposed bypass preventive mechanism node routing discovery process for detection of black hole and greyhole nodes in invincible ad-hoc on demand DV protocol	AODV Protocol (Invincible)	2018
[13]	Here author proposed more secure and reliable data communication in military on fuzzy logic strategy for improving performance of AODV protocol. This strategy is to detect certificate authority, energy handling, packaged veracity checked, and trust node based black-hole attacks	AODV Protocol	2017
[14]	Here author proposed novel strategy for detection (identification) of single and collaboration of black-hole attacks. In this paper D-MBH algorithm is proposed for single and multiple black-hole nodes. Also the author applied additional route request by making use of nonexistent target address, that create a list of abnormal nodes such as black-hole nodes and after than proposed D-CBH algorithm	D-CBH Algorithm& AODVProtocol	2016
[15]	Here author used threshold value of dynamic sequence number and MBDP-AODV protocol found impact of black-hole attack and calculate the performance by parameters are packet delivery ratio, throughput of packet	AODV Protocol	2019
[16]	Here author proposed a MBDF-AODV protocol for performing the detection and prevention of attacks by mitigating black-hole attack type. A dynamic threshold value of the destination sequence number is used for this purpose. This protocol performance validity on the NS-2.35 simulator. After than finally found this proposed protocol is better in performance under black-hole attack	MBDF AODV Protocol	2018
[17]	In this paper author provided comprehensive analysis study about black-hole attack in infrastructure-less network. Here also use CUSUM (Cumulative Sum) test for the detection of the changes of normal behavior of AODV protocol sequence number parameter	AODV Protocol	2017

(continued)

Table 2 (continued)

Ref. No.	Techniques used	Used protocol	Publication year
[18]	The paper proposes a approach for detection and elimination of malicious nodes on the AODV protocol in MANET. In this approach data control packets are utilized to check the nodes in the selected path using extended data routing information table. This approach implemented in different scenarios of OPNET 14 and finally this approach improves the throughput of network	AODV Protocol	2016
[19]	In this paper author proposed a new enhanced secure AODV protocol which is secured by black-hole and grey-hole attacks in MANET. This approach is simulated on NS-2 software and found its more secure methodology compared to current existing one	AODV Protocol	2017
[20]	In this paper, the author proposed new algorithm which had found identify of malicious nodes and after that delete them from routing process. This is new algorithm provides more security of AODV protocol. This algorithm was simulated on NS-2 software and results are good for end-to-end delay and packet delivery rate	AODV Protocol	2016
[21]	The author of this paper proposed an approach for AODV protocol which makes more secure and trusted. Its call by name STAODV that approach work on previous information than it makes isolated incoming packets from malicious nodes	AODV Protocol	2017
[22]	The author in this paper proposed a scheme for identifying malicious nodes. In this scheme, neighbors receive a control sequence and based on these sequences node can expect node response	Black-hole Attack	2015
[23]	Here, the author proposed a technique for detection of black-hole nodes. The defined technique with the help of trap method aims to detect malicious nodes. A black-hole node is detected when OE detection is performed. Also, an alarming method is triggered to alert other nodes for being aware of malicious nodes	AODV Protocol & Black-hole Attack	2016

5 MANETS Major Challenges

Based on the study, following are some crucial challenges observed during the study.

Challenge	Description
Limitation of bandwidth	In MANET each node share common bandwidth and any node don't know other node how much bandwidth consumes so it is also the major challenge
Routing overhead	Routing overhead is exchange different types entries of routing table on the MANET in different exchange time periods
Dynamic topology	In MANET dynamic topology has disturbed the trust full relationship within the networks
Hidden terminal problem	At the receiving node collision of packets when occurring of simultaneous transmission that nodes not in the range of sender node but in the range of receiver node
Transmission error and packet losses	It is also a major challenge in MANET due to overflow of buffer, link failure, no destination route or node found that time data packets to be loss
Security issue by threads	Like Denial of service, eavesdropping, spoofing etc are generate security issues in MANET
Mobility induced route changes	That issue generated on geographical routing related like location error
Tiny battery size	Each node maintains tiny battery for energy. So every node have lightweight with small battery backup and life
Misbehaving nature by selfish nodes	These nodes have not the drop data packets in network that selfish node only stole the some important information related to network or data-like cooperative behavior of nodes, packet delivery ratio, etc they can do further attack
Misbehaving nature by malicious nodes	These nodes directly drop the data packets. These nodes after the receiving of packet does not forwarded to another node

6 Conclusion

In the presented paper, the authors have investigated the DYMO and other routing protocols w.r.t their security issues and attacks. They presented the survey work for the attacks such as black hole and/or warmhole attack in mobile ad-hoc network. As the part of future work, it is attempted to explore more methodologies and solve the issues regarding security of ad-hoc network. The main aim would be to implement an efficient mechanism for handling these attacks.

References

1. D. Nitnaware, A. Thakur, Black hole attack detection and prevention strategy in DYMO for MANET. in *IEEE 3rd International Conference on Signal Processing and Integrated Networks*, pp. 279–284 (2016)

2. E. Zola, I.M. Escalona, DYMO self-forwarding: a simple way for reducing the routing overhead in MANETs. Hindawi Mob. Inf. Syst. **2017**, 10–17 (2017)
3. R. Dutta, R. Thalore, Performance comparison of AODV and DYMO routing protocols, for congestion detection in VANET, in *International Journal of Advance Research Ideas and Innovations in Technology*, pp. 447–454 (2017)
4. J. Kaur, R. Kaur, Performance analysis of AODV and DYMO routing protocols in MANETs using cuckoo search optimization. International Journal of Advance Research in Computer Science and Management Studies **2**, 236–247 (2014)
5. A.K. Gupta, H. Sadawarti, Performance enhancement of DYMO routing protocol with ant colony optimization. Int. J. Electron. Electr. Eng. **2**, 188–193 (2014)
6. S.K. Deb, P.K. Banerjee, Modified dynamic MANET on-demand (DYMO) routing protocol, in *International Journal of Emerging Trends & Technology in Computer Science*, pp. 139–144 (2014)
7. A. Pandey, R. Thalore, Performance comparison of AODV and DYMO routing protocols for boundary detection in 3-D wireless sensor networks, in *International Journal of Advance Research, Ideas and Innovations in Technology*, pp. 429–436 (2017)
8. N. Arya, U. Singh, S. Singh, Detecting and avoiding of worm hole attack and collaborative black hole attack on MANET using trusted AODV routing algorithm, in *IEEE International Conference on Computer, Communication and Control*, pp. 1–6 (2015)
9. A.K. Yadav, A. Kush, Assessment of routing protocols in MANET. IJCSC **07**, 252–257 (2016)
10. D. Nayak, Y.C. Kiran, Malicious node detection by identification of gray and black hole attacks using control packets in MANETs. Imperial J. Interdiscip. Res. **3**, 494–498 (2017)
11. P. Pongle, G. Chavan, Real time intrusion and wormhole attack detection in internet of things. Int. J. Comput. Appl. **121**, 786–797 (2016)
12. V.S. Venu, D. Avula, Invincible AODV to detect black hole and gray hole attacks in mobile ad hoc networks. Int. J. Commun. Syst. **31**, 1–19 (2018)
13. G. Arulkumaran, R.K. Gnanamurthy, Fuzzy trust approach for detecting black hole attack in mobile ad-hoc network. Mob. Netw. Appl. **2017**, 1–8 (2017)
14. K.S. Arathy, C.N. Sminesh, A novel approach for detection of single and collaborative black hole attacks in MANET. Proc. Technol. **25**, 264–271 (2016)
15. S. Gurung, S. Chauhan, A dynamic threshold based algorithm for improving security and performance of AODV under black-hole attack in MANET. Wireless Netw. **25**, 1–11 (2017)
16. S. Gurung, S. Chauhan, A dynamic threshold based approach for mitigating black-hole attack in MANET. Wireless Netw. **24**, 2957–2971 (2018)
17. C. Panos, C. Xenakis, Analysing, quantifying, and detecting the black hole attack in infrastructure-less networks. Comput. Netw. **113**, 94–110 (2017)
18. A. Dorri, An EDRI-based approach for detecting and eliminating cooperative black hole nodes in MANET. Wireless Netw. **23**, 1767–1778
19. S. Dhende, A. Najan, SAODV: black hole and gray hole attack detection protocol in MANETs. (WiSPNET), in *Proceedings of the International Conference on Wireless Communications, Signal Processing and Networking, Chennai, India, 22–24 March*, pp. 2391–2394 (2017)
20. S. Shahabi, M. Bakhtiarian, A modified algorithm to improve security and performance of AODV protocol against black hole attack. Wireless Netw. **22**, 1505–1511 (2016)
21. M.B.M. Kamel, A.N. Onaizah, STAODV: a secure and trust based approach to mitigate black hole attack on AODV based MANET, in *IEEE in Proceedings of the IEEE 2nd Advanced Information Technology, Electronic and Automation Control Conference (IAEAC)*, Chongqing, China, 25–26 March, pp. 1278–1282 (2017)
22. A. Dhakaa, R.S. Dhaka, Gray and black hole attack identification using control packets in MANETs. Proc. Comput. Sci. **54**, 83–91 (2015)
23. N. Sharma, A.S. Bisen, Detection as well as removal of black hole and gray hole attack in MANET, in *Proceedings of the 2016 International Conference on Electrical Electronics and Optimization Techniques (ICEEOT)*, Chennai, India, 3–5 March, pp. 3736–3739 (2016)

Outlier Detection Using Trust Assessment Scheme in Wireless Sensor Networks

Charu Goyal, Vaibhav Aren, and Sarishty Gupta

Abstract With the advancement of technology, one of the major fields of evolution is the growth of Wireless Sensor Networks. The sensors play a vital role in collecting neighbourhood data based on physical distance and spatial–temporal factors. This paper proposes a trust assessment scheme, where each node effectively utilizes neighbourhood data and physical distance of neighbouring nodes. Trust score is calculated for each epoch over 54 different nodes of the Intel Berkeley Lab with weight given to previous sensor data as well. A cyclic framework is proposed to obtain the trust scores that show correlation property which means the trust score of data influences the network node's trust score. This trust assessment scheme is weighted clustering approach, each node shares its data with its neighbours and the effective value of the trust is based on present trust and trust of the previous recorded epoch. K-means algorithm is applied to each node-data which consists of neighbourhood data as well. The algorithm uses X-Control charts statistics to segregate readings as safe or unsafe. The results are compared with another distance-based assessment scheme, where trust sensed value is calculated for each reading, and the trust sensed score is computed for each node. Both algorithms are compared considering the classification accuracy measure, complexity and in terms of memory and energy constraints. The accuracy of the proposed trust assessment scheme is 92.3%.

1 Introduction

Wireless sensor networks (WSNs) are made out of a vast number of nodes that are in close proximity to each other to exchange data [1]. They are designed to perform specific functionalities and are developing points in the field of communication sci-

C. Goyal (✉) · V. Aren · S. Gupta
Department of CSE and IT, Jaypee Institute of Information Technology, Noida, India
e-mail: charu.charu015@gmail.com

V. Aren
e-mail: vaibhavdaren@gmail.com

S. Gupta
e-mail: sarishty.gupta@jiit.ac.in

A. Nanda and N. Chaurasia (eds.), *High Performance Vision Intelligence*, Studies in Computational Intelligence 913,
https://doi.org/10.1007/978-981-15-6844-2_5

ences. Numerous industries including defence, transportation and farming can profit by the positioning and capacities of the wireless sensors. In such new situations, sensor networks gather and store enormous amounts of data and a lot of information that can pass on significant data timely and precisely for decision-making. Each sensor is controlled from a base station. The sensor nodes can be utilized collectively to gather data for the purpose of examining the characteristics of the surroundings where the WSN is deployed. When mote strikes a balance, they may imperil the system by infusing false information. Sensor nodes are subjected to size and cost imperatives which bring about comparing requirements on resources such as energy consumption, memory usage, computational speed and communications data transfer capacity [6].

One of the major necessities of the users of the WSNs is the trust usefulness of the data recorded by the sensors. This requirement prompts a research problem called as outlier detection and trust calculation of each wireless sensor node. The data values whose attributes differ significantly from the standard data patterns depending on a specific measure are declared as outliers. Outlier detection can be utilized to detect corrupt data, identify untrustworthy data patterns and find unique events [2]. For example, sensors can be deployed around borders (remote and harsh areas) that would help in surveillance and detecting any anomalous behaviour or intrusion. These sensors collect a large amount of data and can possibly fail because of energy exhaustion, limited bandwidth and even intrusion attack. A possible solution for resource exhaustion issue is to associate a trust measure with every data node. This trust result will give a sign about the reliability of the sensed node and then subsequently be used for comparison or/and classification of data. Trust value provides a probabilistic measure that informs us about how much the node can be trusted.

The major contribution of this paper is to propose a trust assessment scheme for detecting the outliers in WSNs and contrast it with distance-based trustworthiness assessment scheme [6]. Moreover, the algorithm proposed here has the following properties: (1) it is robust concerning network and data change; (2) the outcome depends on values of all the sensors; (3) it is universal—appropriate for many outliers' detection heuristics. Most trust-based algorithms deployed to detect malicious nodes in WSN take into account the shortcomings of sensor networks in their prototype and plan so that the consumption of energy of mote is reduced and the lifespan of the network system is accelerated keeping up the integrity of the specifications.

Rest of the paper is organized as follows. Section 2 discusses the literature related to the work. Section 3 demonstrates the experimental methodology. Section 4 shows the results of the proposed approach and analyses them. Section 5 describes the conclusion of our research work.

2 Related Work

This section presents the literature covering various approaches designed to measure the trustworthiness of a network. The unit associated with sensors to measure its reliability is called trust. There are varieties of techniques to compute successive node's trust. The techniques include the reputation-based, collaborative, event-based and agent-based schemes for trust management. As discussed by Vitaly and Carolyn [12], the reputation-based scheme, the success of the node determines how trustworthy the node is. Higher the success rate in terms of delivering packets to the neighbouring nodes, more reputed the node would become. License for accessing a protected node/resource is given to nodes with a high reputation. Haiguang Chen Huafeng Wun et al. [13] showed how in the event-based trust management system, at specific time events, the trust is calculated periodically. The node uses an agent to monitor all the events happening is its neighbourhood and subsequently maintains a trust table for all these events. Syed and Jan [14] discussed that in agent-based trust management systems, each agent has its unique features and stores its own trust that is periodically calculated. In this scheme, by aggregating the different trust score of each sensor, a collective trust for the whole system is established. Further classification of spatial approach includes centralized as well as decentralized approaches. A Boukerche and Li [7] proposed that the centralized neighbourhood aggregation requires exhaustive calculation and hence becomes more tedious and time consuming. In the decentralized method of trust calculation, every node maintains its own trust table which is used to assess the trust of its neighbouring nodes.

On the contrary, a centralized approach evaluates the trust score for all the sensors deployed in the network. For carrying out localized calculations such as routing protocols, a decentralized approach would fit in the best. Whilst for carrying out calculations that involve the whole system, a centralized approach will be best suited. Many trust-based schemes involve trust calculation based on the physical distance from their neighbours which is called a distance-based approach. Lim et al. [5] presented a centralized approach that used normal distribution to calculate the trust score for sensors. Closer is the value to the sensed value, higher is the trust associated with it. Zhang et al. [11] discuss a trust-based method for reliable data collection. The mote's trust is calculated using a system metric event. Chen et al. [5] presented the anomaly detection scheme based on spatial and temporal variations. This scheme uses the concept of dividing sensors into various groups based on similarity. In order to detect the anomaly correlation of different nodes is taken into account time-wise. Different groups coordinate with each other to decide whether the anomaly is present or not. This method only is based only on a single attribute correlation and does not consider multivariate attribute correlation.

Al-Zoubi et al. [8] used the approach of machine learning for disaster detection, mainly commercial fires in WSNs for event detection. They used a decision tree and reputation-based trust management scheme for performance evaluation of sensors. In collaborative models, trust collaboration from different agents is taken into consideration. Access control of different agents is necessary to maintain the level of security. In contrast, Al-Zoubi et al. [9] proposed fuzzy logic for outlier detection. For the detection of the remaining outliers, an objective function is first determined. Subsequently, the difference between values calculated from objective function and the frequent change of sensor vales gives the points which are categorized as outliers. Mohamed and Kavitha [10] proposed a live classification technique for outlier detection in WSN. They used Support Vector Machine (SVM) classification technique for categorization of the node as an outlier. The problem in this method is the high complexity of computation because of frequent updates of the sensor values. Classification as a network outlier generally implies an event whereas cluster outlier means it is an error. Ganeriwal et al. [3] presented reputation-based trust management for outlier detection. It takes into consideration the trustworthiness and different type of deviation of nodes values in the network. This technique makes use of Bayesian formulation for trust calculation. Furthermore, Asmaa Fawzya et al. [4] presented a technique of side by side clustering and outlier detection method. In recent times, models for outlier detection are continuously evolving. E-commerce models have grabbed attention for trust management.

3 Experimental Methodology

3.1 Dataset Description

The dataset is collected from the Intel Berkeley Research lab. Figure 1 shows the relative position of 54 sensors, which are deployed in the Intel Research lab. These sensors gather timestamped topology information, alongside temperature, humidity, voltage and light values every 31 s. MoteId ranges from 1 to 54. Temperature is measured in terms of degrees Celsius. Relative humidity has a range varying from 0 to 100%. Voltage is measured in volts, which ranges from 2 to 3 and Light is measured in Lux.

All findings are based on the Intel Berkeley Lab. Different moteId at the same time generate two or more sensor readings at epoch number. There are some missing

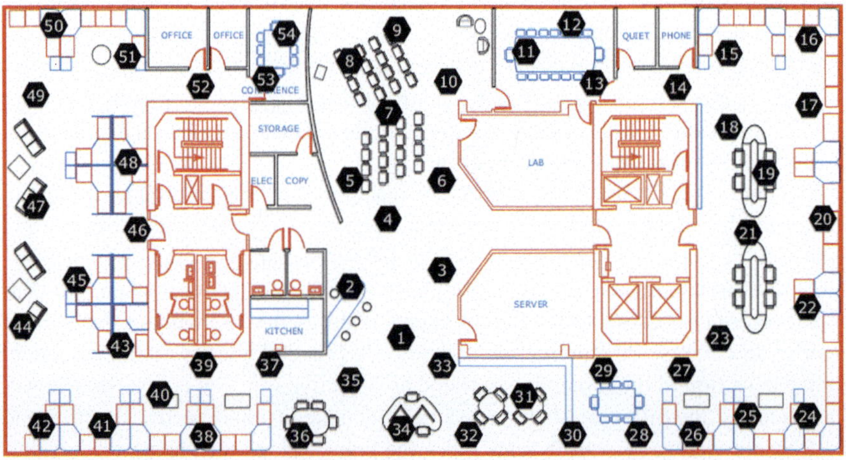

Fig. 1 The position of sensors in Intel Lab

timestamps in this dataset. Range of sensor nodes is from 1 to 54; data from certain motes might be missed or truncated. For the preprocessing step, missing epochs for moteId were ignored. The values which were not in the given range were classified as outliers.

3.2 Proposed Trust Assessment Scheme

In this algorithm, each sensor shares its sensor data with its neighbours within the range of 10 m. Frames of each node with individual and neighbour data are created. The upper control limit and lower control limit are calculated for each mote. The sensor values that do not lie between this range are discarded.

K-means algorithm is applied on each node-data which consists of neighbourhood data as well, and data is divided into 50 clusters and clusters are classified as outliers if average inter-cluster distance is within [mean-std, mean+std] of the mean distance between the clusters. The trust score is calculated for each epoch over 54 different nodes with weight given to previous sensor data as well. To compute trust scores, a periodic system is proposed which well reflects the property of interrelatedness. This implies that the trust score of the sensor nodes in the network is affected by the trust score of the sensor. This trust assessment scheme is a decentralized approach, each node shares its data with its neighbours and the effective value of trust depends on current trust and the trust on the previous recorded epoch. The algorithm uses X-Control charts statistics to segregate readings as safe or unsafe. The algorithm for the proposed trust assessment scheme is as follows:

- Segregate the Intel Berkley Lab data based on various moteIds. Each mote shares its sensor data with its neighbours within the range of 10 m.
- Repeat the following steps for each epoch
- Main two counters for all54 nodes, i.e. safe and unsafe and initialize both the value to 1
- Calculate Lower Control Limit (L.C.L) and the Upper Control Limit (U.C.L) using the following formulas:

$$\bar{x_i} = \frac{\sum_{j=1}^{n} x_{ij}}{n} \tag{1}$$

$$s = \sqrt{\frac{\sum_{i=1}^{N} (x_i - \bar{x})^2}{N - 1}} \tag{2}$$

$$LCL = \tilde{x} - m\left(\frac{d}{\sqrt{n}}\right) \tag{3}$$

$$UCL = \tilde{x} + m\left(\frac{d}{\sqrt{n}}\right) \tag{4}$$

- Consider the value of m to be 3
- Discard the sensor reading that is not within the L.C.L, U.C.L limits.
- Repeat the following steps for each sensor reading considered for X-Control chart:
- If the sensor reading lies between the control limits, increment safe count of the respective node of this reading. Otherwise, increase the unsafe count
- Maximum safe or unsafe count of a particular node can be 54, since the initial value is 1, and maximum neighbours can be 53
- Calculate the trust of each node using values of safe and unsafe readings for each node

$$\frac{safe + 1}{safe + unsafe} \tag{5}$$

- For each epoch, the final value of trust is calculated by giving some weight to the previous trust as well. Trust is recomputed using

$$trust = \alpha \times C.V + (1 - \alpha) \times P \tag{6}$$

3.3 Distance-Based Trustworthiness Assessment

This section provides an overall view of calculating trust by using a distance-based approach and their respective physical distances [6]. The first step involves calculating TSV, i.e. trust score for every sensed value. For calculating TSV, various inputs like previously calculated trust score for every sensor, physical distance between sensors and sensed value provided by the sensor are taken into consideration. The second step involves calculating TSS, i.e. trust score of each sensor using the previous trust score of sensor and previous trust sense value TSV. The algorithm for distance-based trust assessment scheme is as follows:

- Each mote shares its sensor data with its neighbours within the range of 10 m. TSV is calculated for each sensor reading using the following formula:

$$TSV(t)_i = \frac{1}{1 + |r|}, \tau = \frac{\sum_{j=0}^{n} \frac{(v(t)_i - v(t)_j) \times TSS(t-1)_i^s}{d_{ij}^a}}{\sum_{j=0}^{n} \frac{TSS(t-1)_j^s}{d_{ij}^a}} \tag{7}$$

where i represents a particular mote. j is the mote in neighbour list of i, d_{ij} denotes the distance between mote, i and j and are two weighing factors.

- After TSV calculation, TSS is calculated for each node and each sensor reading as well using following formula:

$$TSS(t)_i = w \times TSV(t)_i + (1 - w) \times (t - 1)_i (0 \leq w \leq 1) \tag{8}$$

- The initial value of TSS was set as 0.5.

4 Results and Analysis

The results for our proposed trust management scheme are shown below. Our proposed trust assessment algorithm gave an accuracy of 92.3%. Figure 2 shows the variation in trust scores when the value of w is varied. Higher w indicates a faster evolution of trust scores. It represents how quickly the trust value of the sensor progresses as we move on. It basically signifies how much importance is given to recent evaluated trust scores than the older ones. Greater the value of w means more focus is given to newer values. In our research work, we have analysed our algorithm for trust calculation by taking three different values of w, i.e. w = 0.2, 0.4 and 0.7. Figure3 depicts the count of readings of each sensor node having a trust score less than 0.5. This means that if the lower count of readings is less than 0.5, the more reliable the node is. Figure 4 and 5 represent a variation of trust score for most untrustworthy and most trustworthy node, respectively. From the graph, we can observe that for the most untrustworthy node, graph varies so much between trust values of [0, 1]. On the other hand, for the most trustworthy node, the graph seems to be stable between [0, 1].

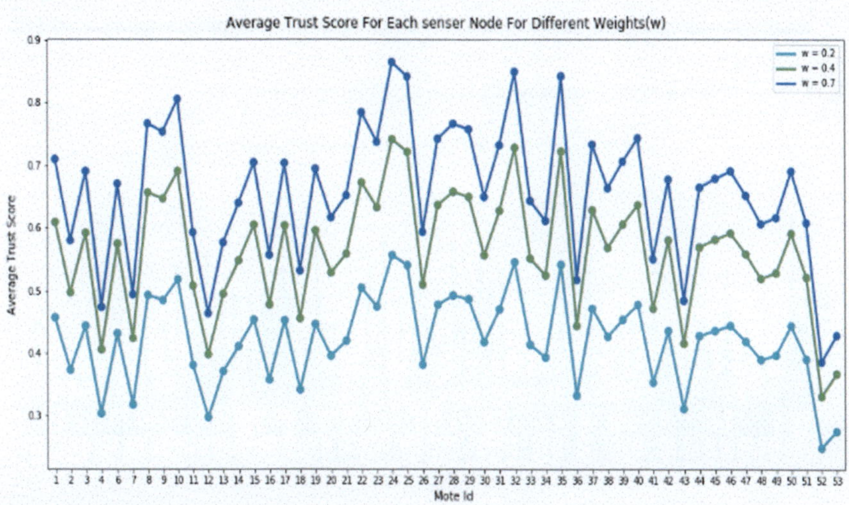

Fig. 2 Average Trust of each sensor varies when w is changed to 0.2, 0.4 and 0.7

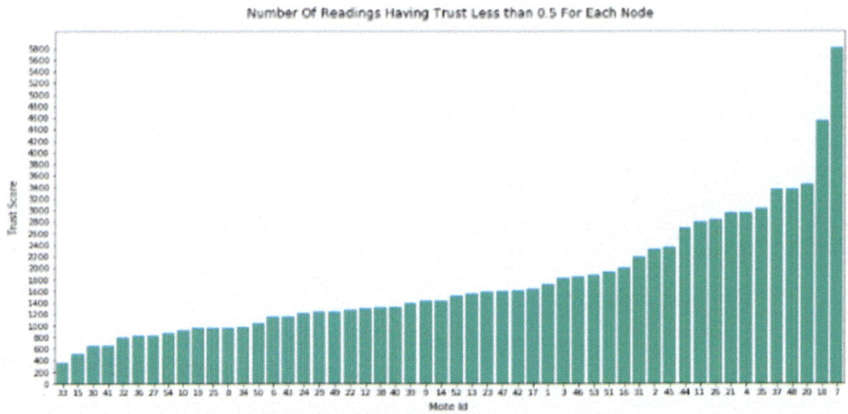

Fig. 3 Count of readings of each sensor node having trust score less than 0.5

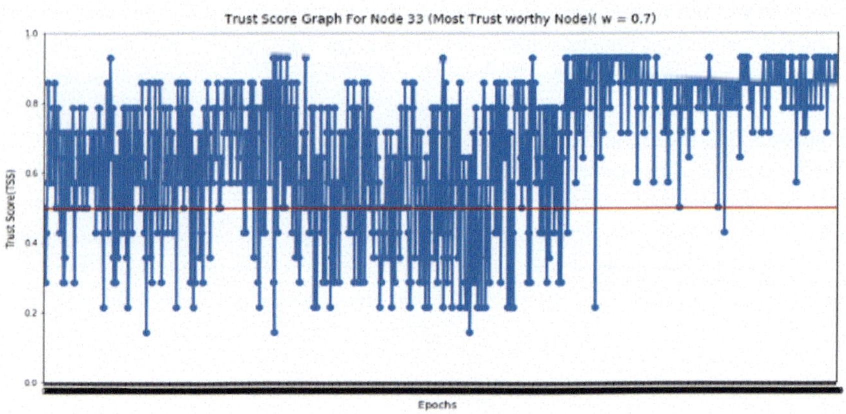

Fig. 4 Trust Score of Node 7 varies which is the most untrustworthy node

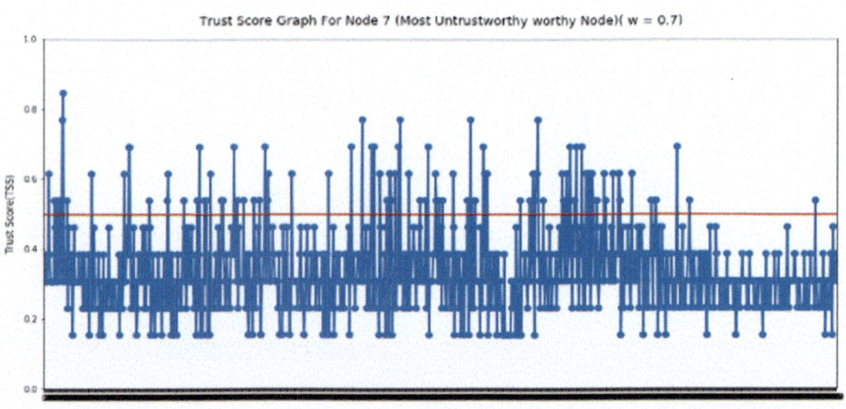

Fig. 5 Trust Score of Node 33 varies which is the most trustworthy node

Table 1 Comparison of Trust Models

Trust mechanism	Method	Trust values	Advantages	Limitations	Complexity	Accuracy
Distance-based approach	Centralized Approach	Uses distance approximation techniques. Computes TSS and TSV for each node	Effectively uses neighbourhood data and physical distances of nodes from its neighbours, Most reliable and scalable	High-energy consumption. Most computational overhead. No temporal correlation	To store past actions of sensor nodes high memory is required	54.71%
Proposed approach	Weighted parameters clustering-based approach	Uses weighted trust calculation. Readings are segregated as safe or unsafe	Energy saving capabilities. Facilitates data aggregation. Scalable for large sensor networks	The trust value depends on cluster heads and it can vary depending upon observations from trusted neighbouring nodes	For ordinary sensor nodes no memory and computational complexity	92.30%

4.1 Comparison of Trust Models

We have compared two trust assessment models, i.e. Distance-based approach and our proposed approach. The comparison of both trust models is presented in detail in Table 1. These models are compared in terms of their methodology of trust calculation, their advantages and limitations and most importantly the computational complexity and accuracy. Whilst distance-based approach uses the physical distance from the neighbours for trust calculation, our approach is a weighted parameterized clustering-based approach. Since our algorithm uses k-means clustering, this algorithm makes it scalable for larger datasets.

For the distance-based approach, calculation of the physical distances of all the neighbours makes the algorithm quite expensive. Our proposed trust assessment scheme gave an accuracy of 92.3%, which is very much higher than that of the distance-based trust assessment model having 54.71%.

5 Conclusion

Trust calculation plays a significant role in determining the degree reliability of other nodes in the network. It is based on the presumption that a reliable node has a high probability of having trustworthy neighbourhood nodes. There has been much research and discussion conducted on trust management scheme of WSNs. In this paper, an attempt was made to propose a novel trust-based algorithm and then subsequently compare and analyse it with the distance-based trust assessment scheme. Based on the results and analysis, both distance-based and novel approach have their own advantages and limitations. Since ours is a large dataset, using a novel approach is more beneficial than the distance-based approach for trust calculation because of energy and memory constraints. High accuracy of 92.3% of the proposed algorithm over distance-based trust assessment scheme also indicates that our proposed approach is more significant for the classification of sensor nodes as trustworthy or untrustworthy. The work can be further extended to investigate the trust values of each node during several instances during which the dataset was collected and also aims to introduce noise to evaluate our framework classifying accuracy, since noise introduced will be labelled to further improve our model.

References

1. Y. Zhang, N. Meratnia, P. Havinga, Outlier detection techniques for wireless sensor networks: a survey department of computer science, University of Twente, P.O. Box 217 7500AE, Enschede, The Netherlands
2. Y.B. Reddy, Trust-based approach in wireless sensor networks using an agent to each cluster. Int. J. Sec. Privacy Trust Manag. (IJSPTM) **1**(1) (2012)
3. S. Ganeriwal, M.B. Srivastava, Reputation-based framework for high integrity sensor networks, in *Proceedings of the 2nd ACM workshop on Security of ad hoc and sensor networks (SASN '04)* (2004), pp. 66–77
4. A. Fawzya, H.M.O. Mokhtarb, O. Hegazy, Outliers detection and classification in wireless sensor networks. Egypt. Inf. J. (2013)
5. H. Chen, H. Wu, J. Hu, C. Gao, Event-based trust framework model in wireless sensor networks, in *International Conference on Networking, Archi-tecture, and Storage* (2008), pp. 359–364
6. Networks Jongho Won(B) and Elisa Bertino, Distance-Based Trustworthiness Assessment for Sensors in Wireless Sensor, International Conference on Net-work and System Security NSS 2015: Network and System Security pp. 18–31
7. A. Boukerche, X. Li, An agent-based trust and reputation manage-ment scheme for wireless sensor networks, in *IEEE GLOBECOMM, 2005.2nd ACM workshop on Security of ad hoc and sensor networks (SASN '04)* (2004), pp. 66–77
8. M. Bahrepour, N. Meratnia, M. Poel, Z. Taghikhaki, P.J.M. Havinga, Distributed event detection in wireless sensor networks for disaster management. Intell. Netw. Collab. Syst. 507–512 (2010)
9. M. Al-Zoubi, A. Al-Dahoud, A.A. Yahya, New outlier detection method based on fuzzy clustering. WSEAS Trans. Inf. Sci. Appl. 681–690 (2010)
10. M.S. Mohamed, T. Kavitha, Outlier detection using support vector machine in wireless sensor network real time data. Int J Soft Comput Eng **1**(2) (2011)
11. Q. Zhang, T. Yu, P. Ning, A framework for identifying compromised nodes in wireless sensor networks. ACM Trans. Inf. Syst. Secur. **11**(3), 12:1–12:37 (2008)

12. V. Shmatikov, C. Talcott, Reputation-based trust management; computer science laboratory SRI international menlo park, CA 94025 USA. J. Comput. Secur. (Special issue on WITS'03 archive) **13**(1), 167–190 (2005)
13. H.C. Huafeng Wu, J. Hu, C. Gao, Event-based trust framework model in wireless sensor networks, in *International Conference on Networking, Architecture, and Storage*. Accessed 28 March 2017
14. S.W. Jaffry, J. Treur, Agent-based and population-based modeling of trust dynamics, in *Conference: 10th International Conference on Social Informatics, SocInfo'18*

An Assessment Model to Establish the Use of Services Resources in a Cloud Computing Scenario

L. Davila Nicanor, H. R. Orozco Aguirre, and V. M. Landassuri Moreno

Abstract When a system or application is designed to run on the Cloud, the scope on storage, users, infrastructure needs, are set by the use, based on the practical environment. It is necessary to replace practices based on experiences and take into account the measurement practices offered by Quality of Service. The main goal of this chapter is to present an assessment model of the availability and efficiency of the existing applications in Cloud Computing to establish their priority of use through statistical simulation and graph models. This requires an analysis of the concerns of the applications available in the Cloud Services. This model is projected as a guide that provides predictive parameters for its evaluation and availability based on the operation of applications from the point of view of user. To create this model, it is necessary to take into account an analysis of the potential risks during the execution of the application and the data analysis query provided to users, in order to efficiently manage resources in an organization.

1 Introduction

Nowadays, everything is moving toward the use of the Cloud, because it is much more productive for any company, that the supplier makes sure of its website, its connectivity, storage capacity, and the availability of the service. Cloud Computing (CC) refers to one set of attributes that any Information Technology (IT) infrastructure implements, which is a feasible alternative for business owners, where information is the basis of the business. Sharing resources represent the biggest benefit of Cloud

L. Davila Nicanor (✉) · H. R. Orozco Aguirre · V. M. Landassuri Moreno
University Center UAEM Valley of Mexico, Autonomous University of Mexico State, Boulevard University, Predio San Javier, Atizapan de Zaragoza, Mexico State, Mexico
e-mail: ldavilan@uaemex.mx

H. R. Orozco Aguirre
e-mail: hrorozcoa@uaemex.mx

V. M. Landassuri Moreno
e-mail: vmlandassurim@uaemex.mx

A. Nanda and N. Chaurasia (eds.), *High Performance Vision Intelligence*,
Studies in Computational Intelligence 913,
https://doi.org/10.1007/978-981-15-6844-2_7

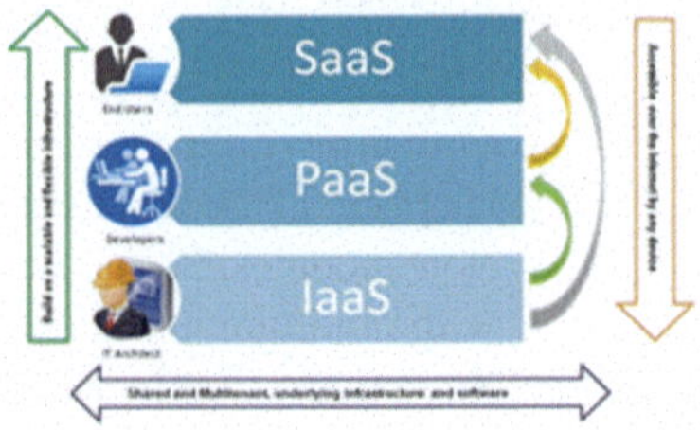

Fig. 1 Cloud computing service taxonomy

Computing Platforms (CCP), because is the way for establishing a means of help for users without worrying about an information technology infrastructure administration. The processing in CC is high performance, which gives this great capabilitie such as: availability, scalability, and efficiency. Service providers also ensure reliability and security, these attributes have been widely discussed by various authors and have been the cause of insecurity in the use of their services for a part of users [1]. Without a doubt, the CC suppliers have purposed to improve these aspects for the benefit of costumers. The key attributes that distinguish CC from traditional computing solutions have been identified in [2–5] and generally comprise the following ones (see Fig. 1):

- Underlying infrastructure and software: These are abstracted and offered as a service, in which service providers usually do not detail the infrastructure characteristics, but they incorporate them in frameworks. In most of cases, the justification for this lack of information has to do with the business level competency scheme. The advantage of the users is focused on a guaranteed service.
- To be shared and multitenant: Each customer of the service is considered a tenant, this allows to customize elements of the application, such as interface colors, but the code is not customized as such, like as the functionality objects in Java Language. It is important to highlight in this architecture that a client is not necessarily a single user, a client can be a group of users.
- To be accessible over Internet: In this case, technologies that are used and promoted in CC guarantee ergonomics in any device such as the mobile ones.

Their capacity stands out before any scheme. The definition of the National Institute of Standards and Technology [7] establishes that "Cloud computing is a model for enabling convenient, on-demand network access to a shared pool of configurable computing resources that can be rapidly, provisioned and released with minimal management effort or service provider interaction". According to [8], the CC categorization is set by the following services:

- Software as a Service (SaaS): This is a level of services, in which the primary feature is that the certain software is delivered by the web platform, generally provided by a web browser to an end user.
- Platform as a Service (PaaS): Here it is possible to obtain right away integrated and coherently setup application and development platforms. A good example is Lightsail [9] of Amazon Web Services (AWS).

- Infrastructure as a Service (IaaS): It is related to services of the type computing workload, such as servers processing and Storage both of them as a Service.

Nowadays the use of CC is based on experimental practice. That is to say, when a system or application is designed to run on the Cloud, the scope on storage, users, infrastructure needs, are set by the use, based on the practical environment. On the contrary, it is necessary to replace practices based on experiences and take into account the measurement practices offered by Quality of Service (QoS). It should be noted that if necessary, the implementation of service computer schemes, require formal quality controls, which implies establishing projections before implementing any type of service.

However, at the moment, the performance specifications of QoS provided by suppliers are not accessible, so users do not have outlines to determine the service' levels on their projects. In order to measure the performance of CC services, service providers support their users with reference schemes, but these are very general, and each service project has a specific purpose which obeys the customer's business goal. These general reference schemas do not take into account the concerns that each project implies in its operation. In recent years, the use of CC is increasing. It is important to develop methodologies to establish lines of QoS measure in function to the purpose for each project aimed to be executed on CC. Considering providers' general reference schemes and each project particular needs, it is possible to obtain better use of the infrastructure provided by the CC.

The approach here presented considers the use of probability of occurrence in concern terms. A concern term offers demand approaches on services and products risk, because it is simple to translate any problematic where there are users and services, or implicit operations, to a risk model where this problem can be represented by logical and mathematical entities. To enhancement the risk approach, the statistical modeling by its nature has proven to be a convenient technique and tool to be applied in any branch of engineering to be able to establish trends in relationships to the variables in turn. In this scenario, the main goal is to develop tools that consider CC as a measurable benefit option of software project management plans. In this case, the evaluation of the utility that CC can offer is done from the point of view of the functionality of the system by establishing the priority of concerns scenarios. Each scenario is set by means the concerns and the probability of occurrence is the parameter to set the service priority. The presented approach is an evolution of the described in [10, 11].

In the methodology development here proposed, a graph technique is used, by mean it is projected the functionality on each system concern, as a set of nodes that are related, at this arise the Concern Dependency Graph (CDG) is development. The graph provides to the logic of the statistical simulation process that simulates the software system operation. At simulation time, the adjustments of the service parameters that the service providers offer are made, to obtain better levels of performance. The performance levels are determined by the statistical data that the simulations yield. Finally, the probability of occurrence is determined when the optimum levels of service have been reached.

This chapter is organized as follows:

- Second section presents the related work as a brief discussion of the works that have been developed around this proposal is presented and too the basis of the probability of occurrence and the recent simulation works are presented.
- Third section is devoted to introduce the used methodology, where the CDG development is establishment as the first step.
- Fourth section is related to present a case study, in which the Conveyor System is presented and on it the used methodology is applied. The simulation results in relation to the levels of service and optimization are presented by graphs and statistical results.
- Finally, conclusions and future work are presented in the last section.

2 Related Work

Today, the infrastructure of CC has a greater scope for any particular company, not in vain governments have made large investments on it. In this case, it is very attractive for any company to leave aside the preoccupation of establishing its own site, which implies having insurance, updating and maintenance schemes in relation to the service and infrastructure. CC guarantees all these needs and also shared and multitenant schemes, which ensures that any user can access the information that their profile establishes, in any region where data exists and from any device, that is, its availability is always high. Another situation that favors the use of its services is the ability to scale them if necessary. The scalability in other scenarios has very high costs of implementation and operation, in CC it can be carried out by the service provider in a transparent way for the user, who is the one that finally benefits or affects their services.

Workload and storage capabilities give to CC the recognition of practitioners and researchers as a valid solution for data storage and processing in both business and scientific computing [12]. The service providers maintain commitments regarding certain attributes such as security and reliability. The strategies are directed to reduce the uncertainties attached to CC. While it is true that CC is a black box, it is understood that having privacy in relation to how the service is provided is part of the competitiveness.

In a world in constant evolution, to know the trends and the possible effects is the mean to set customers' business goals. From this approach, it is necessary to have instruments that support it in the best way.

2.1 Software Reliability Engineering and Concern Concept

In [13], it is proposed the Software Reliability Engineering (SRE) Process to examine the feasibility on services in new software products. The process is constituted

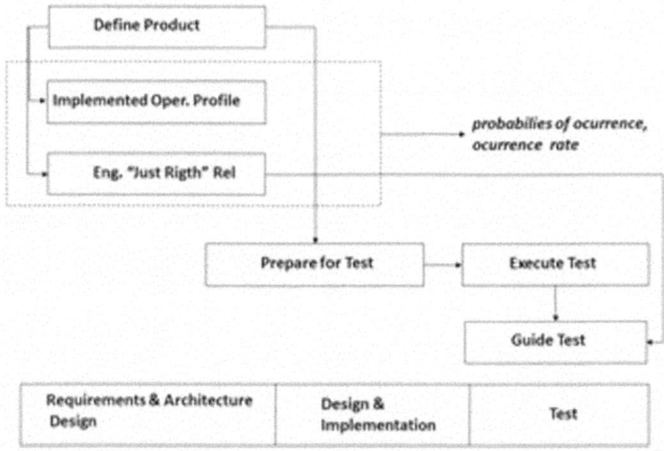

Fig. 2 Block diagram of the SRE process

by six stages and set: the product purpose, functionality, occurrence probability, and occurrence rates in order to software project scope, planning and execution test cases (see Fig. 2). In the first activity the system services are defined: product purpose, supplier and users, and the associated systems are identified. In the operational profiles setting, the functionality is determined by their logical tasks or operations, which is based on its probability of occurrence. The stage is to give information on how users will employ the product and it operation. The operations are the result of the projection of the needs system through the requirements and the design stages. They have a logical relationship and the results of ones are the inputs of others. They can be addressed through the concerns approach. Associated to this, an occurrence rate on an operation is the number of occurrences of the operation divided by the time the total set of operations is running. It is appropriated to evaluate the services of a system in advance. Nevertheless, to evaluate the operation of an application in the Cloud, by means the probability of occurrence in each of the services of the application of interest, in most of providers it is inaccessible. Although determining the availability and efficiency of the service and trends in use in CC is a priority for any interested organization in the use of SaaS, IaaS, or PaaS.

In the functionality concepts, the term concerns set a new vision about the requirements focus, it is possible to construct formal specifications by mean concerns. It has been demonstrated in [14] that the specification of requirements in relation to the software system architecture and the operating environment can be abstracted by the concept of concern through the crosscutting procedure that involves the relationship of specific operations in relation to the software architecture and context operation.

2.2 Cloud Computing Simulators

The simulators are an important tool, these conduce to reduce cost on time and effort, in a software product environment. Another important advantage is that the experiments always have the same operating conditions, which guarantees precision in the results. The simulators use have been applied in fields like distributed computing [15] and grid computing [16], nowadays so in CC context [17–21], in order to simulate workloads running on their operation context. CloudSim simulator [17] is a discrete event simulator, in this the workloads are defined by the users and their instances are submitted and processed by Virtual Machines (VM). In their architecture a scheduler to VM workload has been developed, it is taking into account the space-shared and time-shared according to the availability of processing elements with the goal of optimized the servers use preserving service level agreements existing. The focus of this proposal [18] is to model the workload of the VM at the hardware level. While CloudSim was programmed in Java Language, I CanCloud was made in C ++ Language. Some authors criticize the [17] approach and argue the need to focus the study of the distribution of workloads under random workload schemes [19–21]. The simulator Network CloudSim [19], extends to CloudSim through a network flow model, using bandwidth and latency sharing as parameters, the model takes into account the delays between two directly connected entities. The simulator parameters are configurable in order to set a variety of network topologies. In [20], the CloudSim architecture was extended in terms of energy consumption during workflow execution, by means Dynamic Voltage and Frequency Scaling (DVFS). In [21] the approach is of workload distributions such as servers, switches, and links consumed by the requests of the data center. The scheduler operate by means the server load level and operating frequency.

In [22], the authors argue these simulators are mostly based on application models and simulation algorithms that cannot represent properly the dynamics of the Complex Events Processing (CEP) or Stream Processing systems (SP). The CEP and SP processes are widely used in the Big Data branch, because their operation context improves the latency limitation of other context. In CEP environment the users make query's users (or rules), the analysis is done by means declarative, imperative, and pattern-based languages, Aurora [23] is a Stream Processing (SP) systems based on imperative language. In this context have been proposed StreamCloud [24], Storm [25], and S4 [26]. On complex event processing (CEP) estimation in the Cloud, diverse load conditions have been simulated through the graph technique. In the CEPSim [27] tool, each user queries (or rules) are defined to vertices and the event stream are defined as edges, their relationship generates a Directed Acyclic Graph (DAG), this is the base on analysis. The CEPSim operation is based on processing continuous data flows coming from distributed sources queries mapped into DAGs. The high performance on CC is a necessity and advantage to Big Data research and work. In [28], it presents CEPSim, as a simulator for CEP on cloud environments, on this proposal the treatment of queries users was distributed in two virtual machines and the metrics are latency and throughput.

2.3 Statistical Simulation

This technique is applied to simulate the operation of different productive processes in the real world. The process of interest is usually called a system. This system is developed to study the process from a scientific point of view. In this case, it is needed to take considerations about its operations, these are related through logical and mathematical entities, and their relationships constitute a model that is used to try to acquire knowledge of the environment of the system studied.

If the relationships that make up the model are simple, they can be represented by mathematical methods (such as Algebra, Calculation, or Probability) to obtain accurate information on the issues of interest, this is commonly called an analytical solution. However, most real-world systems are very complex to model with analytical solutions, for their study, it is possible to rely on the simulation context. In a discrete event simulation, computational resources are used to evaluate a numerical model and the data are the inputs to characterize the behavior of the process and estimate the desired data [29]. This technique is a good option to conduct predictability studies in many areas of engineering. In this case it is also for the study of CC operation.

3 Concerns Dependence Analysis

The technique here presented uses steps based on the concern approach and statistical simulation. The probability of occurrence defines the priority of the service on the concern to be analyzed. The separation of concerns principle generally states that it is possible to split different aspects of a problem and deal with each of these subproblems individually [8]. The approach here presented is focused on setting the need of CC services related to system concerns, taking into account the constraints and assumptions related to a system operation.

3.1 Methodology

The proposed methodology has the following four steps:

1. To define the concerns of the application to be analyzed.
2. To Set the CDG based on the concerns of the application define. This graph represents the functionality to specify the normal and special operations required by the user.
3. To perform a statistical simulation based on the operation of the CDG.
4. To select and optimize the parameters and services needs of the Cloud by means the simulation results and to determine the probability of occurrence. Thus occurrence probability are determined as follows:

$$Po\_v_{cn} = \frac{Num\_ocurrences(v_{Ci})}{Total\_No\_occurrences(v_{c1} \ldots v_{cn})}, \tag{1}$$

where $Po\_v_{cn} =$ Probability of occurrence on the particular concern v_{cn} $Num\_ocurrences(v_{Ci}) =$ Num. of occurrence in the particular concern $Total\_No\_occurrences(v_{c1}, \ldots, v_{cn}) =$ Total Num. of occurrence in system concerns

It is important to consider the evaluation of some frameworks on CC, in order to establish which parameters are the most appropriated to be offered by service providers. This is depending on the scope of the system that is intended to be developed using CC services.

3.2 *Dependence Graph*

A CDG is defined by a set of relationships between vertex $V = v_{c1}, v_{c2}, \ldots, v_{cn}$,and edge $E = e_1, e_2, \ldots, e_n$ that represent the service logic in a system, where

1. The sets v, e represent the relationships of concerns system services (SaaS, IaaS, or PaaS service). The edges indicate the direction for each relationship.
2. The first vertex Vin represents the access to CC.
3. The vertices $v_{c1}, v_{c2}, \ldots, v_{cn}$ derived represent associated concern in relation with the services on the CC.

The graph technique is convenient to integrate concerns by means of logical relationships. This was selected because many problematics can be projected by it in a simple and clear way, so the use of this methodology is to set outline the ability to model IaaS and PaaS applications based on real business needs.

4 Case Study: The Conveyor System

Important companies are migrating their work schemes based on CC services. In this case, scaling a service according to the needs is important to ensure the continuity of the service, CC offers this scalability in a transparent way for the user, who can be a provider of accounting or administrative services for example. Having the secure of extending a service in capacity and efficiency without the concern of intermittency represents a great advantage, in addition to the commitment of CC service providers to also have reliable and efficient processes [30].

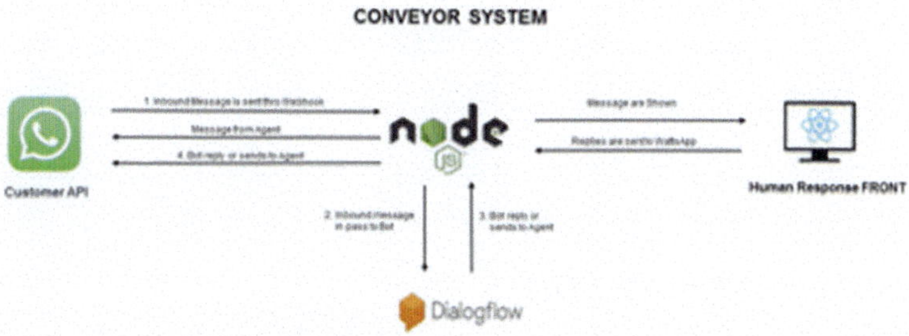

Fig. 3 Conveyor system overview

4.1 Dependence Graph

To evaluate IaaS and PaaS, in a real-study case, the Conveyor system is part of the solution of a Latin American-based sales company that provides orders to their customers with a platform to order different products they need at home, within a certain radius of a location have been taken into account. It has its own application and website with restaurants and shops.

The service given is the result of the relation by means human factor and the Bot technology, they are present in each step of the order. In the process the supplier, the consumer, and the operator are present; in this way, situations arise that are dependent on the platform itself. These problems have to be addressed in the most effective way, the company seeks a more efficient service by getting closer to its customers every day and implementing technologies for it. This is how the Conveyor project was born, which is a system that connects on the one hand to the client by WhatsApp and on the other to the customer service agent helped by the Bot to attend basic requests or campaigns with promotions (see Fig. 3).

This project has a projection in several Latin American countries, such as Brazil, Colombia, and Argentina. The access to the system is by WhatsApp, this application is the most widely used mobile application in several countries, with more than one billion users worldwide. Conveyor system takes this information into account to provide better communication with its users. This will provide a clear competitive advantage for the company. Figure 3 shows the process, the operation system implies the customer arrive by WhatsApp interface on mobile device, the system attends the petition by a graphical interface programmed on NodeJS Language and DialogFlow to the Bot chat. The customer has two ways, the first is the chat between the customer and the Bot. When the conversation is Bot out of bounds, the chat is addressed to customer and human respond front. In the service time, one customer can be made in one order or more, and these can be taken by the system. When the customer have been finished, the departure process is assigned.

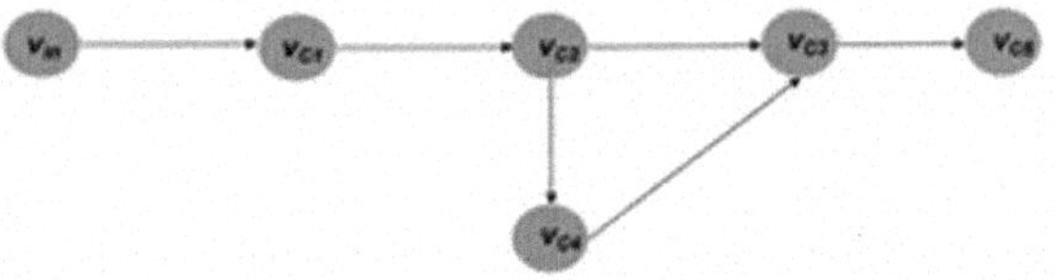

Fig. 4 Concerns Dependence graph (CDG) for IaaS and PaaS to conveyor system

4.2 Conveyor System's CDG

In the development of the CDG, five concerns have been established, in this case the vertices (v_{c1}, v_{c2}, v_{c3}, v_{c4}, and v_{c5}) establish the logic of the relationship of the functionality of the system, thus the vertex v_{c1} represents the arrive state in the system by a WhatsApp customer call, the vertex v_{c2} represents the message from agents on the system in relation to customer active, the vertex v_{c4} is the conversation between the Bot and the customer. The vertex v_{c3} is between human response front and customer, finally, the vertex v_{c5} is the departure's customer action. In Fig. 4, it is possible to see the graph representing the relationships in the Conveyor System. The logical relationships from the operation between PaaS and IaaS concerns are represented by a CDG. This is so important because it is the base to set the statistical simulation. By mean simulation it is possible to set the priority on the resource provisioning in CC.

4.3 Statistical Simulation Based on the CDG Operation

The sequence of the activity of the Conveyor System process is according to the CDG as shown in Fig. 5, by a sequence diagram component test process Java. In this process, customers can access through an App, WhatsApp according to the vertex v_{c1}, nodeJS for the vertex v_{c2}, Dialogflow for the vertex v_{c4}, human response front in the vertex v_{c3}, and the departure system in the vertex v_{c5}. The simulation conditions are similar to the ones reported in [10], the differences are the following ones:

- In the system operation, the work load is projected to 3000 customers ($k = 3000$), this number is based on a project scope. The hosting service used, LightSail AWS, offers 16 GB of memory, 4 core processor a SSD of 320 GB and 6 TB data transfer.
- Depending on the application and coverage of the sale, customer can perform only two types of chats. The operating time depends on the customer-related chat allocated. The Bot respond is faster and limited in the logical than human respond front. To the Bot, the Dialogflow Enterprise Edition are considered because to set up conversations on different devices and platforms.

The sequence between Conveyor System operations is set by sequence diagram showed in Fig. 5. In this process, the customers activity is represented by start, arrive, job, and departure operations at NodeJS, DialogFlow, and Human Respond Front. To initialize the simulation, the *Start_System* instance is activated and data

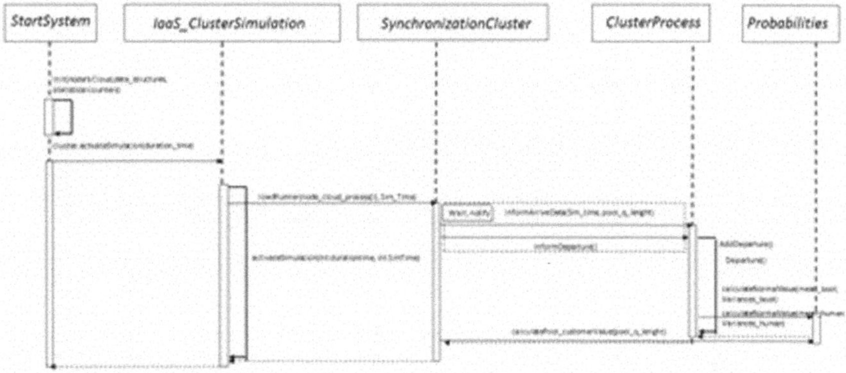

Fig. 5 Cloud simulation to conveyor system

structures and statistical counters at the nodes in each Cloud thread are initialized so the first arrival is scheduled. The main thread simulation is controlled by *IaaS_ClusterSimulation* instance, and the general synchronized nodes on the simulation are controlled by *SynchronizationCluster* instance, when it is the time to the next arrival, it is verified the servers queue capacity and if the system has the capacity to serve the request, it ingresses to the system.

The *ClusterProcess* instance is the process that controls the main operation to the customer service work on the system. The customer operation time (*service_time*) is set by the *AddDeparture*() method, based on a normal distribution according to the average service. In accordance of the customer activity logic and the service time limit, *Bot_service*() and *Human_respond_service*() method are reacted, the Bot's speed is faster than Humans Respond service. When the time customer departure is near to the limit, the *departure*() method is activated by ClusterProcess instance and the event is stored in the back logs. The statistical parameters from customer, orders, and time of service are stored on logs by probabilities instance while the simulations run. Finally, based on statistics, the occurrence probability of concerns is obtained. In order to obtain the frequency to set the occurrence probability, 1000 simulations were executed, the execution time for each simulation was 1000 units of time.

4.4 Probability of Occurrence

In Fig. 6, customer's arrival services have been plotted, on the axes x are programmed customer's arrive and on the axes y are represented the attended customer's arrives. The plot in Fig. 6 shows that the capacities that have been estimated in the CC services are insufficient, because the simulation have programmed a Maximum of 745 arrivals of which the Average reports that only 323 were reached (see Table 1). In this case, the

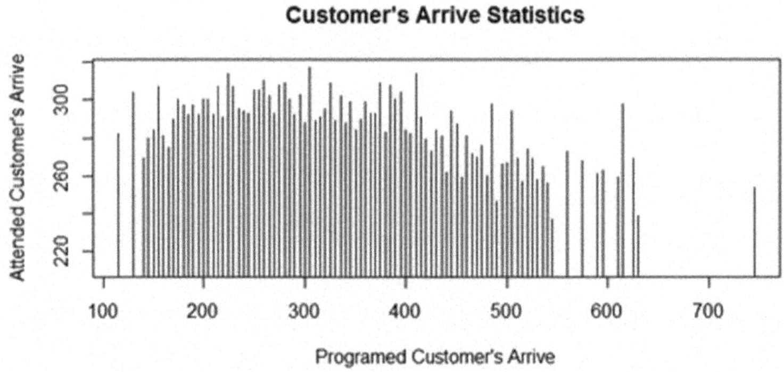

Fig. 6 Customer's arrive on conveyor system statistics

Table 1 Arrives statistics on 1000 simulations

	Programed customer's arrive	Attended customer's arrive
Min.	115.0	211.0
1st Qu.	260.0	256.0
Median	315.0	267.0
Mean	323.1	267.6
Max.	745.0	317.0

system performance is operating below 43.35% of the total capacity of the projected IaaS and PaaS services. In this case, 56.65% of customer requests have been lost, which has a negative impact on the use and dissemination of the system.

In Fig. 7, service Bot time have been plotted, in the x lab are media service Bot time on the y lab are maximum value Bot service Time. The simulation have programmed a Maximum = 46 of maximum value service Bot time, which the Average reports that only 2.319 were used (To see Table 2). To accord these results the Dialogflow Enterprise Edition is correct because their response time is fast in relation of the maximum value.

The extensive benefit of CC is its scalability quality attribute, the change in the workload is imperceptible to the user, and a big advantage is the provisioning of resources to scale it on-demand [30]. The simulation values were adjusted to take into account a more robust infrastructure (see Fig. 8, the hosting service used to Cloud nodes, Lightsail AWS [30] offers 649 GB Memory, 8 core processor 640 GB, SSDs, and 7 TB data transfer. For reasons of reliability in the operation of the system can work properly, where k = 6000 customer.

In Fig. 9, customer's arrival services have been plotted, on the axes x are programmed customer's arrive on the axes y are attended customer's arrive. The plots show that the capacities that have been estimated in the Cloud services are sufficient, in this case the system performance is operating below 81.5% of which the Average

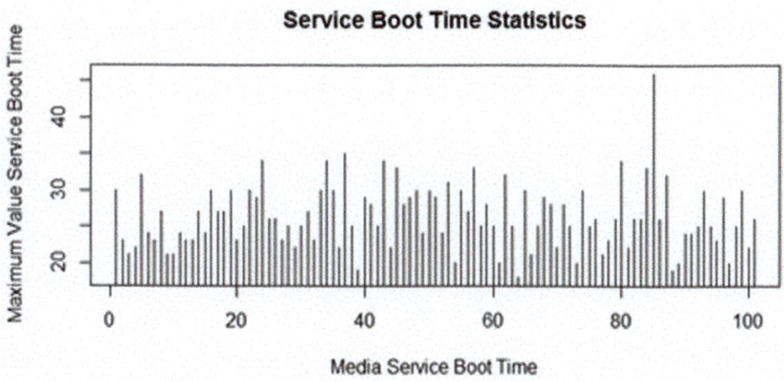

Fig. 7 Service bot time assigned to customer on conveyor system plot

Table 2 Bot time service

	Media service Bot time	Maximum value service Bot time
Min.	2.216	18.00
1st Qu.	2.278	23.00
Median	2.320	26.00
Mean	2.319	26.31
Max.	2.453	46.00

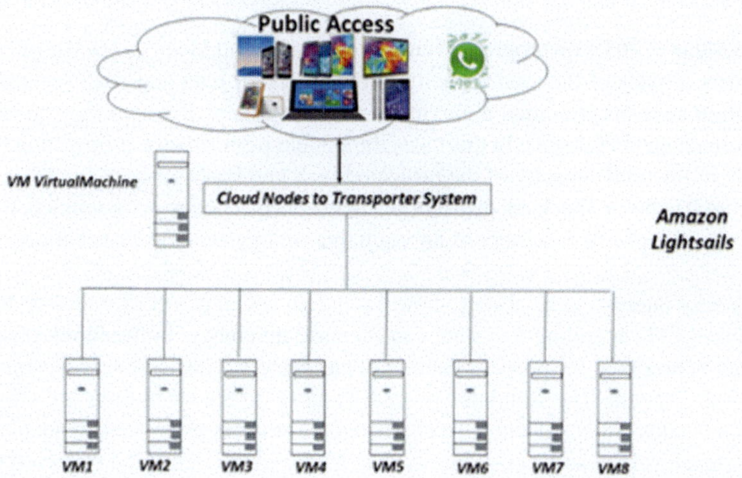

Fig. 8 Cloud simulation IaaS services view with eight virtual machines

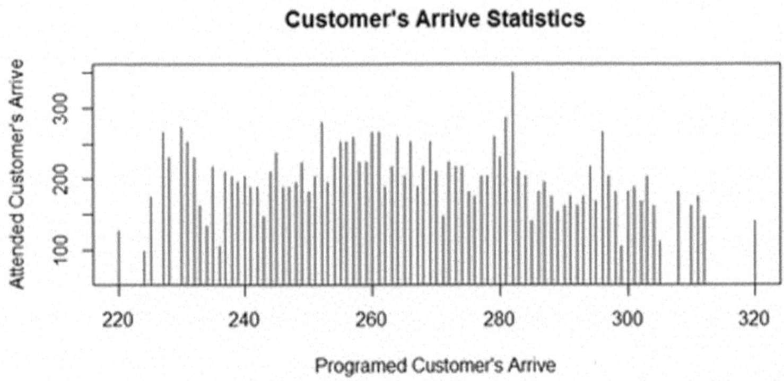

Fig. 9 Customer's arrive on conveyor system statistics with 7 TB data transfer plot

Table 3 Arrives statistics on 1000 simulations

	Programed customer's arrive	Attended customer's arrive
Min.	220.0	63.0
1st Qu.	256.0	112.0
Median	267.0	133.0
Mean	267.2	218.0
Max.	350.0	320.0

reports Mean = 267.2 on programed customer's arrive and Mean = 218.0 on attended customers arrivals of the total capacity of the projected IaaS and PaaS service. The simulation have programmed a maximum of 350 arrivals of which the reports that 320 were reached (Table 3). In this case, the system performance is operating below 91.42% of the total capacity of the projected IaaS and PaaS services.

Figure 10 shows the load distribution between the nodes programmed for the Cloud. In this plot it is observed an equitable and systematic distribution on the nodes.

It is very interesting to observe how having an adequate set of nodes or Virtual Machines (VM) attending a system can increase efficiency. In the development of simulation, however, the proper selection of parameters is decisive for efficient operation of the system. The simulated process is more expensive in time and effort for the Cloud nodes projected when the characteristics that are estimated do not have the appropriate levels to serve customers in the Conveyor System. The observed effect is that in the system pool many clients remain waiting for access, finally only 40% of the requests are answered. On the other hand, when the distributed environment with greater capacity is programmed, times are obviously reduced and we find greater precision in relation to the parameters that are evaluated. In this way, 91.42% of customer attention is achieved, which benefits the system greatly because a system with a low latency is projected.

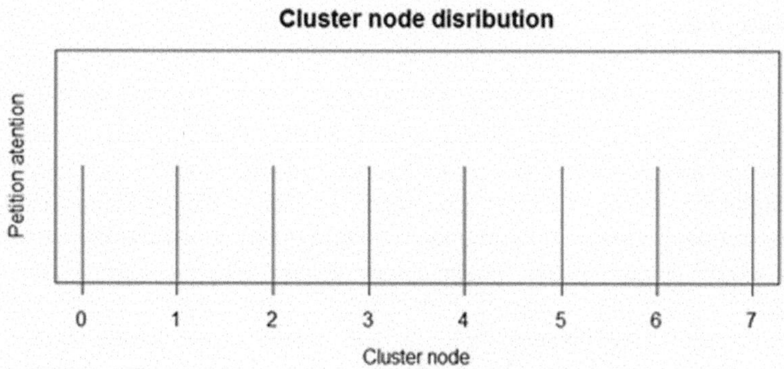

Fig. 10 Cluster node distribution

Table 4 Values obtained with 1000 simulations to CDG of IaaS and PaaS

IaaS and PaaS	Service vertex	Probability of occurrence
WhatsApp access	V_{c1}	0.175
nodeJS (conveyor system support)	V_{c2}	0.239
Human respond front	V_{c3}	0.304
DialogFlow enterprise edition (Bot)	V_{c4}	0.152
Departure system	V_{c2}	0.128

When the selection of parameters is projected in the operation of the simulation, important changes are seen in its processing. In this case, the simulation that emulates a set of nodes with four kernels, gives us a time of 10 min to run 1000 simulations, but when a set of nodes with 8 kernels is projected it is possible to run 1000 simulations in 1.30 min, both executions of the simulation was performed on the same computer equipment, so the memory and processing conditions were the same. The results according to CDG (see Table 4) allow to establish a statistical prediction in relation to values of the system behavior in the CC. The occurrence probability allows establishing criteria and parameters related to latency, availability, and scalability of the evaluated application.

According to the results, it can be observed that the distribution of services is logical and adequate, in relation to the problem evaluated, this is after adjusting the capacities in load storage and the transmission speed. An advantage of this approach is that the projected number of virtual machines in the operating environment depends on the size of the memory where the simulation is run.

This process immediately reduces production costs, due to the adjustment made before real operation, which also guarantees the service and availability for the end user.

5 Conclusions

It is necessary to develop tools that consider CC as a measurable benefit option of software project management plans. The evaluation of the utility that CC can offer, in this proposal is done from the point of view of the functionality of the system by establishing the priority of concerns scenarios. Following the development software process, the specification of requirements in relation to the software system architecture and the operating environment can be abstracted by the concept of concern. Therefore, it is possible to establish that the predictive CDG model on IaaS and PaaS services have the robustness to work with and update service metrics and actualizations related to the analyzed system. The CDG graph controls and optimizes. From concern perspective supported by the logic of the graphs and the predictability of the statistical simulation, generate a robust approach, capable of responding to needs that are difficult to quantify only with the experimental method in the Cloud infrastructure.

The obtained results show that the relation of CDG and statistical prediction allow to establish prediction in relation to values of the system behavior in CC as service time, use frequency, and load work. So the approach here presented allows to establish criteria and parameters related to latency, availability, and scalability of the evaluated application. Finally, the occurrence probability set judgment of service prioritization when the system is optimized. The presented methodology has simple inputs and processes that can be systematized.

6 Future Work

With the obtained data from execution of the case study in the model, it is necessary to systematize the methodology and to determine the direct and indirect metrics that can be obtained. In the future, it is so important to development a CC simulator with the use of agents and statistical approach, in order to have the possibility to predict the use of system applications where it is necessary to test for availability, define critical paths and analyze the functional relationships of the whole system by layered CC taxonomy.

Acknowledgements This work was supported by the Quality on Software and High Performance Computer Laboratories of University Center UAEM Valley of Mexico of the Autonomous University of Mexico State.

References

1. S. Mathur, Moving to cloud computing can be smart decision for governments & companies. The Economic Times **7**, 18 (2013)

2. M. Armbrust, A. Fox, R. Griffith, A.D. Joseph, R.H. Katz, A. Konwinski, G. Lee, D.A. Patterson, A. Rabkin, I. Stoica, M. Zaharia, Above the clouds: a Berkeley view of cloud computing. Technical Report UCB/EECS-2009-28 (EECS Department, University of California at Berkeley, 2009)
3. H. Li, C. Spence, R. Armstrong, R. Godfrey, R. Schneider, J. Smith, R. White, Intel cloud computing taxonomy and ecosystem analysis. IT-Intel Brief (Cloud Computing) (2010)
4. T.H. Oh, S. Lim, Y.B. Choi, K.R. Park, H. Lee, H. Choi, State of the art of network security perspectives in cloud computing, in *Security enriched urban computing and smart grid*, ed. by T. Kim, A. Stoica, R.S. Chang. Communications in computer and information science, vol. 78 (Springer, Berlin, 2010), pp. 629–637
5. Wang, L., von Laszewski, G., Younge, A. et al. Cloud computing:a perspective study. New Gener. Comput. (2010) 28: 137. https://doi.org/10.1007/s00354-008-0081-5
6. Q. Zhang, L. Cheng, R. Boutaba, Cloud computing: state-of-the-art and research chal-lenges. J. Internet Serv. Appl. **1**(1), 7–18 (2010)
7. National Institute of Standards and Technology, *The NIST Definition of Cloud Computing*, http://www.nist.gov/itl/cloud/upload/cloud-def-v15.pdf
8. V. Oliver, A. Ingo, C. Arif, K. Timo *Software Architecture* (Springer Nature, New York, 2011)
9. https://aws.amazon.com/es/lightsail/
10. A. Benhumea-Pena, L. Davila-Nicanor et al., Predictive model to determine quality of service on cloud computing: service dependence graph SDG, in *13th IEEE International Conference on Networking, Sensing and Control (ICNSC'2016), Mexico City, Mexico* (2016)
11. L. Davila, H. Orozco, Modelo de analisis para evaluar la respuesta del servicio en el diseno de las Arquitecturas de las Tecnologias de la Informacion. Coloquio de Investigacion Multidisciplinaria **2**(1), 12 (2014)
12. A. Tchernykh, S. Uwe, A. Vassil, T. El-ghazail, Towards understanding uncertainty in cloud computing resource provisioning. Procedia Comput. Sci. (2015)
13. J.D. Musa, *Software Reliability Engineering: More Reliable Software Faster and Cheaper*, 2nd edn. (AuthorHouse, Bloomington, 2004)
14. R. Clark, A. Moreira, Constructing formal specifications from informal requirements, in *Software Technology and Engineering Practice* (IEEE Computer Society Press, 1997), pp. 68–75
15. D. Luckham et al., A language and toolset for simulation of distributed systems by partial orderings of events. Technical Report CSL-TR-96-705 (Stanford University, 1996)
16. R. Buyya, M. Murshed, GridSim: a toolkit for the modeling and simulation of distributed resource management and scheduling for Grid computing. Concurr. Comput. Pract. Exp. **14**(13–15), 1175–1220 (2002)
17. R.N. Calheiros, R. Ranjan, A. Beloglazov, C.A.F. De Rose, R. Buyya, CloudSim: a toolkit for modeling and simulation of cloud computing environments and evaluation of resource provisioning algorithms. Softw. Pract. Exp. **41**(1), 23–50 (2011)
18. A. Nunez, J.L. Vazquez-Poletti, A.C. Caminero, G.G. Castane, J. Carretero, I.M. Llorente, iCanCloud: a flexible and scalable cloud infrastructure simulator. J. Grid Comput. **10**(6), 185–209 (2012)
19. S.K. Garg, R. Buyya, Network CloudSim: modelling parallel applications in cloud simulations, in *2011 4th IEEE International Conference on Utility and Cloud Computing* (IEEE, 2011), pp. 105–113
20. T. Guerout, T. Monteil, G. Da Costa, R. Calheiros, R. Buyya, M. Alexandru, Energy-aware simulation with DVFS. Simul. Model. Pract. Theory **39**(2013), 76–91 (2013)
21. D. Kliazovich, P. Bouvry, S.U. Khan, GreenCloud: a packet-level simulator of energy-aware cloud computing data centers. J. Supercomput. **62**(3), 1263–1283 (2012)
22. W.A. Higashino, C. Eichler, M.A.M. Capretz, T. Monteil, M.B.F. De Toledo, P. Stolf, Query analyzer and manager for complex event processing as a service, in *2014 IEEE 23rd International WETICE Conference* (2014), pp. 107–109
23. D.J. Abadi, D. Carney, U. Cetintemel, M. Cherniack, C. Convey, S. Lee, M. Stonebraker, N. Tatbul, S. Zdonik, Aurora: a new model and architecture for data stream management. VLDB J. **12**(2), 120–139 (2003)

24. V. Gulisano, R. Jimenez-Peris, M. Patino-Martinez, C. Soriente, P. Valduriez, StreamCloud: an elastic and scalable data streaming system. IEEE Trans. Parallel Distrib. Syst. **23**(12), 2351–2365 (2012)
25. Storm, distributed and fault-tolerant realtime computation. http://storm-project.net/
26. L. Neumeyer, B. Robbins, A. Nair, A. Kesari, S4: distributed stream computing platform, in *2010 IEEE International Conference on Data Mining Workshops* (IEEE, 2010), pp. 170–177
27. W.A. Higashino, M.A.M. Capretz, L.F. Bittencourt, CEPSim: a simulator for cloud-based complex event processing, in *2015 IEEE International Congress on Big Data* (IEEE, 2015), pp. 182–190
28. A. Wilson, W.A. Higashino, A.M. Miriam, A. Capretz, F. Luiz, B. Bittencourt, CEPSim: modelling and simulation of complex event processing systems in cloud environments. Futur. Gener. Comput. Syst. **65**, 122–139 (2016)
29. W.D. Kelton, A.M. Law, *Simulation Modeling and Analysis* (McGraw Hill, New York, 2000)
30. S. Basu, Cloud augmentation, in *Real World Windows 8 Development* (Apress (Springer), Berkeley, 2013)

Video Synopsis: A Systematic Review

Ansuman Mahapatra and Pankaj K. Sa

Abstract Automated analysis of surveillance video is necessary due to its redundant and uninteresting content. This has led to a rapidly evolving research area called video synopsis or otherwise called as rapid recap system. As the name suggests, it can highly condense a lengthy input surveillance video by extracting and shifting the moving objects in temporal axis and then displays them simultaneously in an non-overlapping fashion. This article investigates the prior arts for synopsis video generation for both single and multi-camera network. This article focuses on providing a brief introduction to the area of video synopsis by discussing the state-of-the-art techniques. Then the end of the article also focuses on a new trend in synopsis generation methods applied to the challenging area of multi-camera networks. This literature identifies various optimization techniques, dataset, and evaluation matrix used for each approach.

1 Introduction

Low cost cameras, high-speed data transmission rate, improved compression techniques, and enormous low-cost storage facilities lead to an overwhelming growth of digital video content. The Internet has additionally supercharged this increment in digital video era. There are three main problems, which will grow with the rate of growth of the video content—storage, communication, and video content analysis. Though we have enormous amount of storage and high-speed communication facilities, analyzing those gigantic video contents manually by a human is impossible

---

A. Mahapatra (✉)
Department of Computer Science and Engineering, National Institute of Technology, Karaikal, Puducherry 609609, India
e-mail: ansuman.mahapatra@nitpy.ac.in

P. K. Sa
Department of Computer Science and Engineering, National Institute of Technology, Rourkela 769008, Odisha, India
e-mail: pankajksa@nitrkl.ac.in

A. Nanda and N. Chaurasia (eds.), *High Performance Vision Intelligence*, Studies in Computational Intelligence 913,
https://doi.org/10.1007/978-981-15-6844-2_8

due to the limitation of time and lack of interestingness in the surveillance videos. Automated computational methods are required to facilitate efficient organization, searching, analysis, manipulation, and representation of this huge volume of acquired video data. A video summary can effectively represents the original lengthy video for easier understanding and representation. A video summary should be considerably shorter than the original video covering maximum information from the original video without containing redundant information.

In the last few decades, video summarization is the area of focus for many researchers. Numerous summary generation schemes has been proposed by various researchers for single camera network. Some important area of research in video summary generation are keyframe extraction or video abstraction [19], video carving [4], video condensation [6], video digest [24], video skimming [20], video summarization [2], and video synopsis [27]. Static and dynamic video summary generation methods are two broad areas of research based on type of summary output. Key frames are extracted and displayed in a storyboard in static summary generation schemes. However, the dynamic video summary generates a short video sequence after extracting relevant frames. The dynamic summaries are more useful as they deliver important information preserving the dynamics of the input video.

There are two popular methods to generate dynamic video summary: Video summarization and synopsis. In video summarization, the set of frames having importance are preserved in the summary. However, video synopsis displays couple of objects simultaneously from different intervals of time. The object's spatial locations are not changed and importance of objects or frames are not considered [28]. The difference between summarization and synopsis is pictorially depicted in Fig. 1. In case of video summarization important segments of the original video is included in the summary, whereas in video synopsis the black and brown persons are simultaneously displayed achieving more compact representation. Many video summary generation

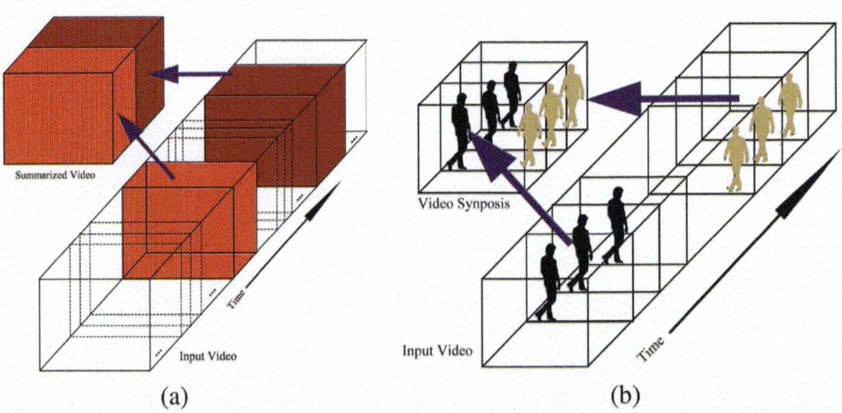

Fig. 1 **a** Important set of frames from the input video is included in video summarization ignoring the trivial contents. **b** Displaying multiple objects simultaneously from different intervals of time in video synopsis ignoring their temporal occurrence

techniques can easily discard the inactive frames temporally, but video synopsis can be able to fit objects from different intervals of time in spatially inactive areas, achieving shorter summary. Video summarization is frame-based method where importance is computed for each frame, whereas video synopsis is object-based method where moving objects are extracted and displayed simultaneously by temporal shifting. Video synopsis helps in efficient object searching and retrieval, faster understanding of video content, improved video browsing, and temporal searching of events from large video archives. It can achieve very compact summary by preserving the dynamic effect of the video.

1.1 Introduction to Video Synopsis System

A mono-view video synopsis generation system consists of two phases: an online phase, that captures video from a camera, detect the moving objects in each frame, and stores their tracks in a database for future query and a response phase, that takes the observer's query, generates a concise synopsis by temporally rearrange the tracks by minimizing overlapping of tracks, synopsis length, and displaying maximum objects from original video and represents it before observer for visualization. A generic video synopsis system has basically four major modules:

1. **Video acquisition module** requires a static video camera to capture video with a static background. The acquisition setup plays a crucial role on the performance of the below modules. For instance, it should be placed in such that it can capture objects with minimum occlusion.
2. **Object detection module** detects all the moving objects in the field-of-view of the camera. Object detection algorithm statistically learned scene background. The input test image is compared and background is subtracted to locate pixels outside of the learned pixel value range and are designated as foreground pixels. Tracks of each objects are stored for synopsis generation and visualization phase.
3. **Synopsis generation module** employs an optimization technique to generate the dynamic video synopsis on demand by user. The module shifts the stored tracks temporally keeping the spatial locations fixed and displays multiple objects simultaneously. The optimization function handles the objectives to minimize occlusion among objects, activity loss, and long video synopsis.
4. **Synopsis visualization module** accepts the query from user and displays the generated video synopsis. Figure 3 illustrates an example of video synopsis vizualization generated by BriefCam [1]. The extracted moving objects are displayed over the static background with the original timestamp of their activities. Figure 2 illustrates different modules of an mono-view video synopsis system.

Last decade has seen quite an upsurge in video summary production which is primarily driven by the demands of security establishments. Thanks to the technological leap in video analytics that has helped in making a significant breakthrough in video summary generation. Figure 4 presents a broad classification of video summary gen-

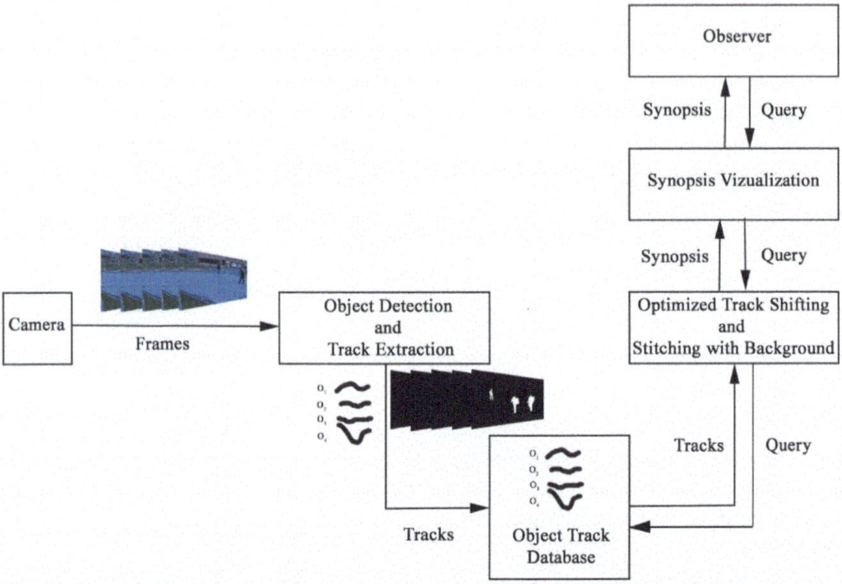

Fig. 2 Different modules of an mono-view dynamic video synopsis system

Fig. 3 A frame from BriefCam [1] video synopsis system

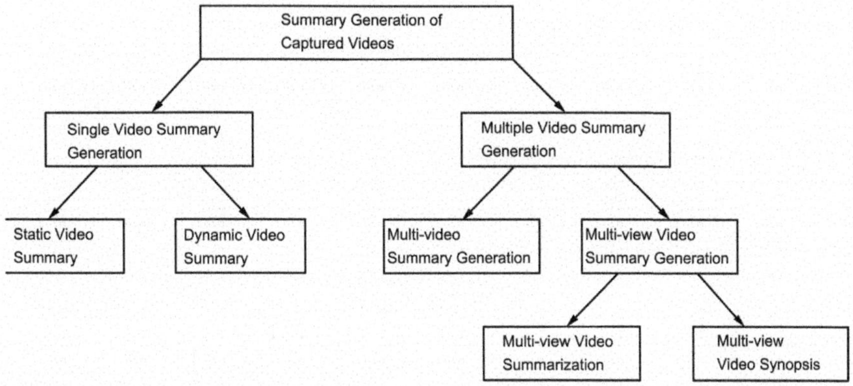

Fig. 4 Classification of various summary generation methods

eration methods based on the number of input videos and type of output summary. Among these methods this article focuses on detailed review of single (described in Sect. 2) as well as multi-view video synopsis (detailed in Sect. 4) techniques. There is a very good review on synopsis techniques by Baskurt and Samit [3]. Readers are highly recomended to have a read on that review article as well.

2 Mono-View Video Synopsis System

Video synopsis technique was introduced by Rav-Acha et al. for a single camera network in which from the surveillance video the objects are extracted and then temporally shifted then displayed simultaneously over a static background, without changing their spatial locations [28]. The tracks of the objects are cut into pieces and displayed in same frames by stroboscopic effect to achieve further reduction in the synopsis length. A synopsis video is generated by optimizing a multi-objective function that maximizes number of objects displayed simultaneously and minimizes the overlapping between objects and synopsis length. The proposed method does not deal with changing background and temporal ordering of the objects' appearance.

Pritch et al. have generated a synopsis video for live surveillance camera using 3D Markov Random Field [27]. The moving objects are stitched with a time-lapsed background to generate the synopsis by minimizing collision between tracks, including more activity, maintaining the temporal order. Yildiz et al. proposed a real-time video synopsis by removing fewer motion areas from video volume using dynamic programming that does not require complex function optimization [34]. Hence, it works faster than the previous methods.

The background of the synopsis video is very dynamic due to change in illumination or object motion. Therefore, two online principal background creation methods has been proposed by Feng et al. [7]. An adaptive background is generated that changes with illumination over time. In another synopsis generation method, Nie et al. have removed the constraint of not changing spatial location. The objects are now shifted both in the temporal as well as spatial axes [23]. Using a global spatiotemporal optimization method activities coverage is maximized by minimizing the spatial deviation, and collision. The clustered synopsis technique proposed by Pritch et al. displays only similar activities to improve visualization and accuracy [25].

A key observation selection-based video synopsis method has been proposed by Zhu et al. that eliminates content-based redundancy [37]. As objects in adjacent frames may appear the same, the method eliminates this replicated observations through time. Significant actions and shape changes are regarded as key observations. To create a more shorter synopsis video a multiple kernel-based similarity method is employed to choose key observations. Choudhary and Tiwari have used video synopsis to index the objects, and the media player plays the corresponding original video when clicked on an object [5]. A contrast context histogram descriptor-based object reappearance model is used by Huang et al. to know already tracked objects [13]. A synopsis table is used which keeps track of each pixel last synopsis time slot in the previous frame. This helps in creating a synopsis video in real time. Xu et al. have proposed a GA-based trajectory combination method and show that it is efficient than simulated annealing in optimizing the information loss and computation time consumption [32]. Pritch et al. have proposed a real-time video synopsis on an endless webcam video. A synopsis is generated based on the query from a user [26].

Lai et al. proposed a method to spatiotemporal track formation method using optical flow and pixel clustering [16]. The arrangement of tracks is done in a Tetris-like strategy by projecting them into X-T, Y-T, and X-Y planes. It keeps the chronological ordering of the objects unaltered. Jin et al. have used a projection matrix to project each dynamically generated tubes to a condensed location [14]. A synopsis method is presented by Vural and Akgul based on operator's eye-gaze [31]. For fast reviewing, the synopsis is generated from the region of interest of monitored videos. A projection matrix is created by projecting each column of the frames vertically. Then an energy matrix is created from this projected matrix. Video ribbon carving is used to remove the 2D sheets from 3D space to condense in the time axis, as discussed in video carving. Dual buffers used to make this method work in real time. Xu et al. proposed a start-time programming method which combines optimal elements in a set [33]. A distributed processing framework for video synopsis is proposed by Lin et al. to speed up computing performance [18]. The median of color is applied to make the foreground object and background seamlessly stitched. Fu et al. proposed a synopsis method in which the interaction between two objects is preserved [8]. The structure is measured by motion proximity and intersection.

He et al. [10] have proposed a graph coloring-based synopsis generation method. Potential collision among the 3D volume of different objects is analyzed geometrically. A graph is formed where each node represents one object, and the edge between them represents the potential collision between them. The advantage of this method is that the collision between objects is calculated once. Kasamwattanarote et al. proposed a direct shift collision detection based synopsis method in which they represent the extracted tunnels in terms of a depth map style which contains the slice number and a flag to check overlapping between tunnels [15]. Unlike most of the synopsis method, this method can work in real time. A just-in-time renderer is used to display the synopsis in real time. He et al. employed a potential collision graph to find the collision between trajectories of objects after projected on a spatial plane [11]. This method can create a synopsis of real-time video by rearranging the tubes in a deterministic way, avoiding complex optimization. The collision between objects can be reduced by replacing the object with a smaller instance of itself while keeping the centroid same [17]. The optimization framework proposed by Li et al. includes both temporal position of an object and resize coefficient. Sun et al. proposed to change the spatial positions of the objects to minimize the collision [30]. Synopsis clips are generated by shifting all tracks to the start, and no further temporal shifting is needed for a clip which reduces the computation time and also decreases the temporal disorder.

In recent work, Gandhi and Ratanpara have used genetic algorithm for efficient tube shifting that preserves the temporal arrangements of the objects in the synopsis [9]. Ghatak et al. in their work, improved the slow convergence rate of the simulated annealing by improving the computation time of energy minimization using hybrid approach using Simulated Annealing and Teaching Learning-based optimization. Ruan et al. have proposed a graph-based track rearrangement technique for online video [29]. The graph nodes are inserted dynamically based on the appearance of the objects in the scene. Then graph coloring algorithm is applied for the rearrangement of the tubes.

Table 1 summarizes the survey on single video synopsis techniques in terms of optimization, datasets, and evaluation metrics used.

Table 1 Summary of mono-view video synopsis techniques (cont.)

Authors	Objectives	Optimization	Datasets	Evaluation metrics
Pritch et al. [27]	Region-based synopsis: Minimize loss in activity, discontinuity cost object-based synopsis: minimize activity cost, collision cost, temporal consistency cost	Graph cut	video streams downloaded from the Internet	Computational cost

(continued)

Table 1 (continued)

Authors	Objectives	Optimization	Datasets	Evaluation metrics
Rav-Acha et al. [28]	Low level approach: Minimize Activity loss and discontinuity cost, object based approach: minimize activity loss, occlusion between objects, long synopsis	Iterative graph cut and Simulated Annealing	Video from a camera monitoring a security fence	✗
Nie et al. [23]	Collision cost, spatiotemporal consistency, activity cost, spatial distance cost	Alpha-beta swap graph cuts method	Five videos	User study based on six different questions
Pritch et al. [25]	Minimize the spatial separation between the activities and maximize the temporal overlap between the activities. Minimize over all temporal shifts, motion distance, and collision cost	Approximate nearest neighbor searching	Video having 100 objects	Length of the synopsis video
Zhu et al. [37]	Minimize cost of stitching object to time lapsed background, loss of key observation, collision cost, and time consistency cost which preserves the chronological order	Simulated Annealing	Two surveillance video dataset and a author's own dataset	Compression rate
Huang et al. [5]	Appearance model, motion model, and the synopsis table is updated for each new object	Graph Cut	Two outdoor and one indoor videos	Ratio of frame reduction and computation time

(continued)

Table 1 (continued)

Authors	Objectives	Optimization	Datasets	Evaluation metrics
Xu et al. [32]	Lost area of objects, overlapping area of objects	Genetic Algorithm	One video from the web and another video from CAVIAR dataset	Information loss and time consumption
Pritch et al. [26]	Minimize activity cost, temporal consistency cost, collision cost	Simulated Annealing	Webcam video streams from the Internet	✗
Lai et al. [16]	Minimize overlapping object tracks	No optimization needed as objects are shifted in X-T plane	Five Video sequences	Length compression and the amount of activity pixels retained, and average percentage of a single frame
Jin et al. [14]	Minimize the sharing of pixels between more than two objects in one frame	No optimization, simple projection matrix is used	Four videos	Speed of number of frames processed, chronological disorder, and condensation ratio
Vural and Akgul [31]	Minimum energy paths removed based on seam carving	Dynamic programming based optimization	An input video where two persons walk	Evaluated based on the operator overlooks any actions
Xu et al. [33]	Minimize the empty space without any visual objects	Combinatorial optimization by continuous relaxation	Two video streams (hall and car parking)	Ratio of the number of pixels occupied by objects to the total number of pixels
Lin et al. [18]	Loss in activity and discontinuity across seams	Greedy optimization	Twelve surveillance video from the "Safety City" project in China	Time compress ratio, capacity compress ratio, and computing time-consuming ratio

(continued)

Table 1 (continued)

Authors	Objectives	Optimization	Datasets	Evaluation metrics
Fu et al. [8]	Minimize spatial collision and temporal consistency	Stepwise optimization method as in [6] to reduce search space	Six videos from public datasets and videos recorded by authors	Condensation rate, and by user study based on the criteria of visual pleasing, compactness, comprehensibility, and overall satisfaction
He et al. [10]	Minimize collision among objects	L(q)-coloring for partitioning the graph	Nine surveillance video sequences taken from indoor and outdoor scenes	Frame condensation ratio, frame compact rate, and overlap ratio
Kasamwattanarote et al. [15]	Minimize spatial collision among tunnels	Dynamic programming	Trecvid London Gatwick surveillance video	Compression ratio
Li et al. [17]	Minimize activity cost, temporal consistency cost, collision cost, and size reduction cost	Simulated Annealing	Four surveillance videos	Sum of all overlapping areas to measure its collisions, collision frames, run time
Sun et al. [30]	Minimize activity cost, collision cost, spatial location cost	A predict cost is used for selecting a tubelet to be included in the synopsis	Public surveillance video	Activity preserve rate, object preserve rate, object conflict rate, condense rate, and chronological disorder

3 Challenges in Multi-view Video Synopsis

Nowadays, multi-camera networks are being predominantly in use as surveillance infrastructure. In such setup, a scene is captured from several viewing angles for better scene understanding. Though synopsis generation seems to be a trivial task at first glance, indeed it is very challenging for various reasons. The following inherent challenges are faced by the researchers while generating synopsis for multi-camera infrastructure.

- **Content redundancy:** In recent past there have been many reported methods for single video summary generation. However, their usage in multi-view videos are limited due to the ignorance of inter-video relations that leads to a summary video with redundant contents. Same objects are appearing in multiple cameras. Therefore, if mono-view video synopsis is applied to each video, then the generated synopsis will contain same objects in all synopsis which leads to redundant content.
- **Difficulty in comprehension:** Within a short span of time synopsis displayed many objects. It is very difficult for the observer to remember the objects which are already appeared in the previous synopsis videos. Therefore, if we generate the synopsis separately for each camera, associating the information among multiple synopsis videos is a critical task for the observer. Therefore, the only solution is to display a single synopsis by combining all the information from all camera views.
- **Multiple synopsis:** The extracted moving objects from the input video frames are represented on a background frame for video synopsis visualization. However, each video in a multi-camera network has separate background. Therefore, if the mono-view synopsis methods are applied they generate multiple number of synopsis.
- **Satisfying observer's multiple demands:** Based on the task at hand, the generated synopsis should meet the variable demands of the observer. The synopsis generation methods should provide the flexibility to the observer in tweaking the parameters based on the demand. Therefore, generating a synopsis by accommodating important objects while minimizing number of collisions and length is a challenging task. In the presence of multiple cameras, the task is further compounded.

4 Multi-view Video Synopsis

Video summarization extracts features from the whole frame. Therefore, they can work with asynchronous video capturing cameras. However, the shot-based summarization methods sometimes yield unusual result as the object track may break during shot segmentation for a continuous traffic surveillance site. Therefore, video synopsis technique is extended to the multi-camera network. A proposition by Zhu et al. extracts object's tracks from each view to do multi-camera joint video synopsis [36]. The proposed optimization rearrange and join the tracks in such a way that the temporal appearance of the object's in all the synopsis remains same. Occlusion cost and chronological disorder cost is minimized by a greedy algorithm. The method can be applied to both partially overlapped and non-overlapping multi-camera network. However, the only disadvantage is that the number of synopsis video generated is equal to the number of cameras. With increase in number of cameras, the task of the observer will be completed in comprehending these synopsis.

A live video synopsis method has been proposed by Hoshen and Peleg [12]. When a person is detected in the live master camera, all the activities captured by one or more slave cameras about that person are displayed to the operator in a video synopsis.

This method suffers from the same limitation in generating number of synopsis video that is equal to the number of slave cameras. Multiple synopsis videos are generated because the background of each view is different and an object cannot be displayed on any common background.

For partially overlapped MCN Zhu et al. have proposed a synopsis generation method [37]. The object tracks are associated with a mosaic plane using a graph matching algorithm. Multiple kernels are defined based on the change in object's appearance, motion, and spatial uniformity to remove the redundancy in an object track. Simulated Annealing is used to minimize a similar traditional multi-objective energy function to generate the synopsis. The extracted objects are stitched to the mosaic plane for the visualization. Only one synopsis video is generated by this method. The authors have considered a MCN with a special type of arrangements of cameras in which the cameras are placed side-by-side and are capturing almost parallel to each other. Therefore, the mosaic looks good in the final synopsis. However, with an increase in camera network and with a change in Field-of-Views of cameras, the mosaic plane may leave holes due to insufficient information provided to the mosaic algorithm and they will not look good while synopsis visualization on that type of plane. Also, the method has not discussed the way to represent an object if there is a significant change in appearance due to FoVs change.

A multi-view video synopsis framework has been proposed by Mahapatra et al. which takes multiple multi-view videos as input and a single synopsis video as output unlike other proposed methods [21, 22]. Object tracks from all the views are mapped on to the top view of the site under surveillance using Homography technique. Objects are plotted as circles and at the bottom, action performed by each object is displayed. They have compared multiple optimization methods in terms of length reduction.

Zhang et al. have proposed a video synopsis solution for multi-camera network, that generates a single synopsis video [35]. The method generates multiple synopses from different camera views by considering three factors; Objects should not move out of frame, minimum overlapping, and maintain chronological order. Then they have applied view selection method to display only one synopsis by selecting that synopsis video which has minimum object overlap and smooth switching. Both objectives are then unified and the multi-objective optimization is solved using graph cut and dynamic programming.

A the state-of-the-art multi-video synopsis schemes are summarized in Table 2. The synopsis generation methods are summarized on five different parameters. The first parameter includes the information about the consideration of objects' importance while generating the summary. The number of summary video generated by each synopsis method is summarized under the heading number of synopsis. The last parameter lists whether best view is considered for synopsis generation.

Table 2 Summary of multi-view video synopsis techniques

Parameters → Approaches ↓	Object's importance	Number of summary	Best view selection
Zhu et al. [37]	✗	1	✗
Zhu et al. [36]	✗	n	✗
Hoshen and Peleg [12]	✗	n	✗
Mahapatra [21, 22]	✓	1	✗
Zhang et al. [35]	✓	1	✓

5 Conclusion and Future Work

This review of existing literature has offered new insights into the domain of video synopsis. Some key observations are as follows:

- Mono-view video synopsis is popular and a great degree of work has been done in this area since a decade.
- However, Multi-view video synopsis is a relatively less explored area. The current synopsis methods generate multiple synopsis videos whose number is equivalent to the number cameras in a multi-camera network.
- Keeping track of objects in multiple synopsis videos is a cumbersome task when the number of cameras grows in a network. A proper visualization technique is required for synopsis presentation before users.
- As moving objects are important components of a surveillance video their importance is considered by few multi-view video synopsis techniques.
- There few areas where enhancement is possible in terms of optimization in synopsis generation and visualization of the objects.

References

1. Briefcam: The Video Synopsis Company. http://briefcam.com
2. M. Ajmal, M.H. Ashraf, M. Shakir, Y. Abbas, F.A. Shah, Video summarization: techniques and classification, in *Computer Vision and Graphics* (Springer, 2012), pp. 1–13
3. K.B. Baskurt, R. Samet, Video synopsis: a survey. Comput. Vis. Image Underst. **181**, 26–38 (2019)
4. B. Chen, P. Sen, Video carving. Short Papers Proceedings of Eurographics (2008), pp. 68–73
5. V. Choudhary, A.K. Tiwari, Surveillance video synopsis, in *Indian Conference on Computer Vision, Graphics & Image Processing*. IEEE (2008), pp. 207–212
6. S. Feng, Z. Lei, D. Yi, S.Z Li, Online content-aware video condensation, in *Computer Vision and Pattern Recognition*. IEEE (2012), pp. 2082–2087
7. S. Feng, S. Liao, Z. Yuan, S.Z. Li, Online principal background selection for video synopsis, in *IEEE International Conference on Pattern Recognition* (2010), pp. 17–20
8. W. Fu, J. Wang, L. Gui, H. Lu, S. Ma, Online video synopsis of structured motion. Neurocomputing **135**, 155–162 (2014)

9. S. Gandhi, T.V. Ratanpara, Object-based surveillance video synopsis using genetic algorithm, in *Censorship, Surveillance, and Privacy: Concepts, Methodologies, Tools, and Applications*. IGI Global (2019), pp. 857–883
10. Y. He, C. Gao, N. Sang, Z. Qu, J. Han, Graph coloring based surveillance video synopsis. Neurocomputing **225**, 64–79 (2017)
11. Y. He, Z. Qu, C. Gao, N. Sang, Fast online video synopsis based on potential collision graph. IEEE Signal Process. Lett. **24**(1), 22–26 (2017)
12. Y. Hoshen, S. Peleg, Live video synopsis for multiple cameras, in *International Conference on Image Processing*. IEEE (2015), pp. 212–216
13. C.R. Huang, H.C. Chen, P.C. Chung, Online surveillance video synopsis, in *International Symposium on Circuits and Systems*. IEEE (2012), pp. 1843–1846
14. J. Jin, F. Liu, Z. Gan, Z. Cui, Online video synopsis method through simple tube projection strategy, in *Wireless Communications & Signal Processing*. IEEE (2016), pp. 1–5
15. S. Kasamwattanarote, N. Cooharojananone, S. Satoh, R. Lipikorn, *Real Time Tunnel Based Video Summarization Using Direct Shift Collision Detection* (Springer, Berlin, 2010), pp. 136–147
16. P.K. Lai, M. Décombas, K. Moutet, R. Laganière, Video summarization of surveillance cameras, in *Advanced Video and Signal Based Surveillance*. IEEE (2016), pp. 286–294
17. X. Li, Z. Wang, X. Lu, Surveillance video synopsis via scaling down objects. IEEE Trans. Image Process. **25**(2), 740–755 (2016)
18. L. Lin, W. Lin, W. Xiao, S. Huang, An optimized video synopsis algorithm and its distributed processing model. Soft Computing (2015), pp. 1–13
19. T. Liu, H.J. Zhang, F. Qi, A novel video key-frame-extraction algorithm based on perceived motion energy model. IEEE Trans. Circuits Syst. Video Technol. **13**(10), 1006–1013 (2003)
20. Y.F. Ma, H.J. Zhang, A model of motion attention for video skimming, in *International Conference on Image Processing*. IEEE (2002), pp. 129–132
21. A. Mahapatra, P.K. Sa, B. Majhi, A multi-view video synopsis framework, in *International Conference on Image Processing*. IEEE (2015), pp. 1260–1264
22. A. Mahapatra, P.K. Sa, B. Majhi, S. Padhy, MVS: a multi-view video synopsis framework. Signal Process.: Image Commun **42**, 31–44 (2016)
23. Y. Nie, C. Xiao, H. Sun, P. Li, Compact video synopsis via global spatiotemporal optimization. IEEE Trans. Vis. Comput. Graph. **19**(10), 1664–1676 (2013)
24. Y. Pritch, S. Ratovitch, A. Hendel, S. Peleg, Clustered synopsis of surveillance video, in *Advanced Video and Signal Based Surveillance*. IEEE (2009), pp. 195–200
25. Y. Pritch, A. Rav-Acha, A. Gutman, S. Peleg, Webcam synopsis: peeking around the world, in *International Conference on Computer Vision*. IEEE (2007), pp. 1–8
26. Y. Pritch, A. Rav-Acha, S. Peleg, Nonchronological video synopsis and indexing. IEEE Trans. Pattern Anal. Mach. Intell. **30**(11), 1971–1984 (2008)
27. A. Rav-Acha, Y. Pritch, S. Peleg, Making a long video short: dynamic video synopsis, in *Conference on Computer Vision and Pattern Recognition*. IEEE (2006), pp. 435–441
28. T. Ruan, S. Wei, J. Li, Y. Zhao, Rearranging online tubes for streaming video synopsis: a dynamic graph coloring approach. IEEE Trans. Image Process. (2019)
29. L. Sun, J. Xing, H. Ai, S. Lao, A tracking based fast online complete video synopsis approach, in *International Conference on Pattern Recognition*. IEEE (2012), pp. 1956–1959
30. U. Vural, Y. Akgul, A parallel non-linear surveillance video synopsis system with operator Eye-Gaze Input. INTECH Open Access Publisher (2011)
31. L. Xu, H. Liu, X. Yan, S. Liao, X. Zhang, Optimization method for trajectory combination in surveillance video synopsis based on genetic algorithm. J. Ambient Intell. Human. Comput. **6**(5), 623–633 (2015)
32. M. Xu, S.Z. Li, B. Li, X.T. Yuan, S.M. Xiang, A set theoretical method for video synopsis, in *Multimedia information retrieval*. ACM (2008), pp. 366–370
33. A. Yildiz, A. Ozgur, Y.S. Akgul, Fast non-linear video synopsis, in *International Symposium on Computer and Information Sciences*. IEEE (2008), pp. 1–6

34. Z. Zhang, Y. Nie, H. Sun, Q. Zhang, Q. Lai, G. Li, M. Xiao, Multi-view video synopsis via simultaneous object-shifting and view-switching optimization. IEEE Trans. Image Process. (2019)
35. J. Zhu, S. Liao, S.Z. Li, Multi-camera joint video synopsis. IEEE Trans. Circuits Syst. Video Technol. **26**(6), 1058–1069 (2016)
36. X. Zhu, J. Liu, J. Wang, H. Lu, Key observation selection for effective video synopsis, in *International Conference on Pattern Recognition*. IEEE (2012), pp. 2528–2531
37. X. Zhu, J. Liu, J. Wang, H. Lu, Key observation selection-based effective video synopsis for camera network. Mach. Vis. Appl. **25**(1), 145–157 (2014)

360° User-Generated Videos: Current Research and Future Trends

S. Priyadharshini and Ansuman Mahapatra

Abstract The 360° video, also known as immersive or spherical video, allows the observer to have a 360° view and an immersive experience of the surroundings. Each direction in this video is recorded at the same time either by an omni-direction camera or by an assembly of cameras synchronized together. The viewing perspectives are controlled by the viewer during playbacks. This article gives an overview of the existing research areas and methods in the user-generated 360° videos for streaming, transcoding, viewport-based projections, video standardization, and summarization. This survey also provides an analysis of the experience estimation in 360° videos. The study of multiple quality evaluation criteria is also reviewed. Moreover, 360° video user experience studies are also focused on this survey. The merits and demerits of each technique are investigated in depth.

1 Introduction

Cameras are affordable these days due to the technology advancements, which leads to a significant utilization of cameras by the users for capturing precious moments in their life. Omni-direction cameras can capture the whole scene using more than one camera and the images captured by these cameras are stitched together to give a 360° view of the scene.

Figure 1 portrays the basic workflow of 360° video. It generally commences with an omni-direction camera capturing 360° frames. Those are organized (i.e., stitched) together and sent to the encoding phase where the spherical video is projected to a 2D plane followed by frame packing and compression. The commonly used two projection formats: Equirectangular Projection (ERP) and Cubemap Projection (CMP) of

S. Priyadharshini · A. Mahapatra (✉)
Department of Computer Science and Engineering, National Institute of Technology, Puducherry, India
e-mail: ansuman.mahapatra@nitpy.ac.in

S. Priyadharshini
e-mail: priya81818@gmail.com

A. Nanda and N. Chaurasia (eds.), *High Performance Vision Intelligence*,
Studies in Computational Intelligence 913,
https://doi.org/10.1007/978-981-15-6844-2_9

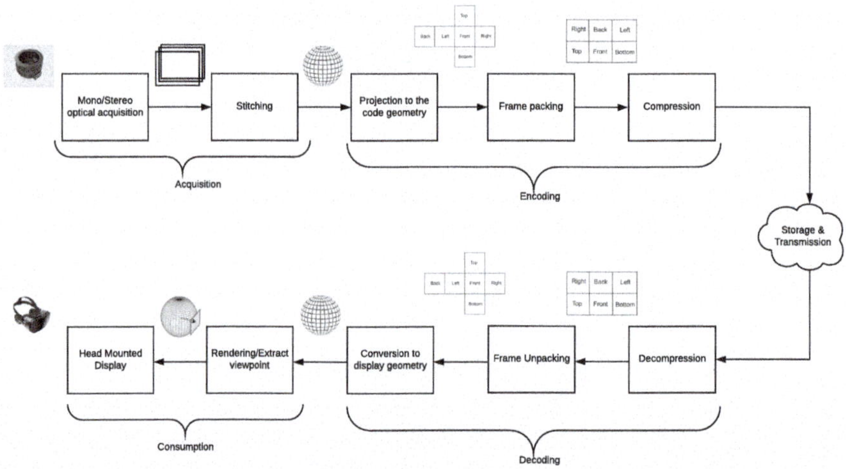

Fig. 1 360° video processing workflow [1]

Fig. 2 Equirectangular projection

a user-generated 360° video are shown in Figs. 2 and 3, respectively. The encoding phase is followed by the decoding phase where a single video undergoes interactive projection that offers the rendering process inter-relating with the respective input/output technology (such as HMD) at the consumer end.

Figure 4 depicts the different FoVs in traditional viewing mode extracted from the equirectangular projection given in Fig. 2. This gives the content creators flexibility to shoot in 360° and later in the post-processing they can select the FoV that matters the most.

This review article

- is the first review on the user-generated 360° video to the best of our knowledge.
- introduces various research areas in the user-generated 360° video.
- investigates recent literature and categorizes based on research areas.
- highlights the pros and cons of each methodology.

The article is organized as follows. Section 2 briefs various research trends in 360° video production, communication, and analysis. The processing techniques applied

(a) Layout (b) An illustration

Fig. 3 Cubemap projection

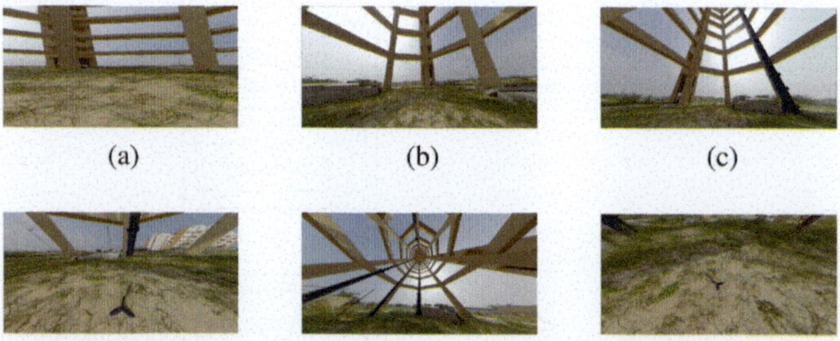

(a) (b) (c) (d) (e) (f)

Fig. 4 **a–f** Different FoVs from user-generated 360° video

on 360° videos are discussed in Sect. 2.1. Section 2.2 discusses steaming techniques. Video post production methodologies are discussed in Sect. 2.3. The evaluation of the quality of 360° videos are reviewd in Sect. 2.4. Observations are listed in Sect. 3 and Sect. 4 concludes this article.

2 Research Trends in 360° Video

A brief survey of each research area in a 360° video is discussed in this section. Figure 5 depicts the research trends in 360° video.

2.1 Processing of 360° Video

This section discusses various processing techniques required for 360° videos before transmitting or storing. After capturing a 360° video, they need to be stitched and projected into a suitable representation, and then it will be compressed for transmis-

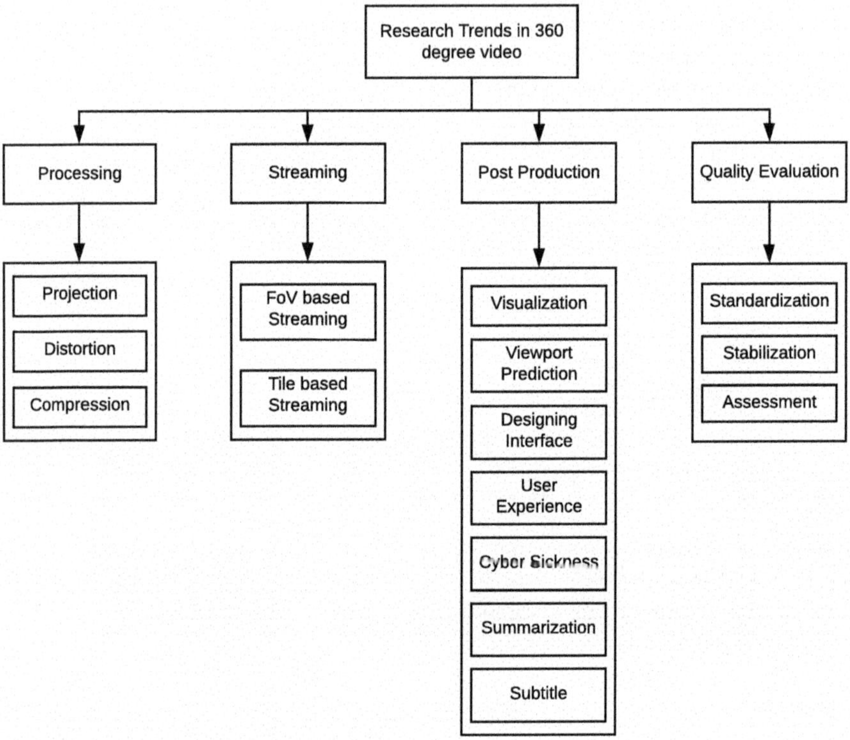

Fig. 5 Areas of research in 360° video

sion or storage. The following subsections present a review of the existing methods in processing 360° video.

2.1.1 Projection

In Sphere Segmented Projection, the visual artifact is caused due to inactive region [2]. In order to enhance coding efficiencies and to minimize visual artifacts, Yoon et al. suggest a scheme of padding inactive region. For panoramic videos, Huang et al. presented a low-complexity prototype scheme and video stitching mechanism [3]. Hanhart et al. recommended a coded approach on the basis of spherical neighboring relationship and projection form adaptation [4]. Su and Grauman proposed a spherical convolutional network used to process 360° imagery straightforward in its equirectangular projection, which is translated from a planar Convolutional Neural Network (CNN) [5]. Lin et al. propose a hybrid equiangular cubemap projection that minimizes seam artifacts [6]. Some characteristic equirectangular projection forms of sequences in the clip are experimented by Wang et al. [7].

It is unfavorable to attain a well-organized compression for storing and transmitting [8]. Hence, Vishwanath et al. recommended a rotational model for identifying the angular motion on the sphere effectively. In 3D space, for an angle α, vector A is rotated around an axis given by a unit vector B. The coordinates of vectors A and B are (p, q, r) and (l, m, n), respectively. The coordinates of the rotated vector $A^{'}$ will be

$$p^{'} = l(B \cdot A)(1 - \cos\alpha) + p\cos\alpha + (-nq + Ar)sin\alpha \tag{1}$$

$$q^{'} = m(B \cdot A)(1 - \cos\alpha) + q\cos\alpha + (np - lr)sin\alpha \tag{2}$$

$$r^{'} = n(B \cdot A)(1 - \cos\alpha) + r\cos\alpha + (-Ap + lq)sin\alpha \tag{3}$$

where $B \cdot A$ is the dot product. Rotation of axis B is the vector right angled to the plane well defined through the origin, vector A, and also rotated vector $A^{'}$. Vector B is computed as follows:

$$B = \frac{A \times A^{'}}{|A \times A^{'}|} \tag{4}$$

Angle of rotation is given as

$$\alpha = \cos^{-1}(A.A^{'}) \tag{5}$$

The summary of techniques, highlights, and challenges of 360° video projections is listed in Table 1.

2.1.2 Distortion

Azevedo et al. provide an extensive analysis of the most popular visual deformity that alters the 360° video signals in immersive applications [1]. Aksu et al. present a scalable multicast live deliver of 360° video with distortion analysis [9]. Yoon et al. recommend an approach of adding inactive regions to lessen deformations [2]. A detailed review of the distortions in 360° video is given in Table 2.

2.1.3 Compression

Le et al. designed a transcoding system with ARIA block cipher for encoding purpose [10]. To gain high steady sampling, Lin et al. offer 360° specific coding tools [6]. The mapping function is given as follows:

Table 1 Summary on projection of 360° video

Author	Technique	Highlights	Challenges
Yoon et al. [2]	Method of padding inactive Regions	Significantly lessens the visual artifact	–
Huang et al. [3]	Procedure for video stitching	Minimizes computation	Finding the best seam has a single constraint
Hanhart et al. [4]	Approach on projection forms	Reduces the face seam artifacts	Dependent on the frame packing configuration
Su and Grauman [5]	Spherical convolutional network	Yields accurate results	Further exploration of spherical convolution is required
Lin et al. [6]	Hybrid equiangular cubemap projection scheme	Achieves increased uniform sampling	–
Wang et al. [7]	Sphere-shaped Coordinates Transform Placed Mobility Method (SCTMM)	Saves significant bits for video sequence	Complexity computation is very high
Vishwanath et al. [8]	Prototype intended for rotational movement	Globally suitable for all the projection geometries	–

Table 2 Summary on distortion of 360° video

Author	Technique	Highlights	Challenges
Azevedo et al. [1]	Investigation on visual distortions	Detects the reasons for deformations	–
Aksu et al. [9]	Distortion analysis	Aggregate distortion is minimized	Decrease in encoding gain is not focussed
Yoon et al. [2]	Method of padding inactive regions	Visual artifacts was minimized	–

2D (Cube-Map) to Sphere:

$$h_b(a, b) = \frac{b}{1 + 0.4(1 - a^2)(1 - b^2)} \tag{6}$$

Sphere to 2D (Cube-Map):

$$k_b(a, b) = \begin{cases} b, & \text{if s=0} \\ \frac{1-\sqrt{1-4s(b-s)}}{2s}, & \text{otherwise} \end{cases} \tag{7}$$

$$\text{where } s = 0.4b(k_b(a)^2 - 1).$$

To improve the quantity of storing and compressing video based on perception, an efficient compression mechanism called Vignette was suggested by Mazumdar et al. [11]. Xiu et al. recommended that gain for the paired categories of video such as HDR and SDR attains considerable efficiency on coding [12]. Aimed at the spherical environment, Wang et al. propose an algorithm for compensation and estimation based on the motion prototype [7].

In order to enhance efficiency and minimize encoding time, Zhang et al. present an optimization procedure on compression [13]. Choi et al. offer an inventive video compression approach for video service accompanied by high quality and video coding schemes using HDR [14]. Lin et al. propose a subject labeled database with a massive scale, which comprises compressed H.265/HEVC videos consisting of miscellaneous PEAs [15]. In order to have an enhanced performance, Le et al. designed a transcoding system that plays a vital role in modifying bit rates and changing the resolution of 360° videos [10]. Various 360° video compression techniques, highlights, and challenges are summarized in Table 3.

2.2 *Streaming of 360° Video*

This section presents various mechanisms required for 360° video streaming. Streaming can be done based on FoV or Tiles. The following subsection gives a detailed description of the techniques involved in FoV-based and Tile-based streaming.

2.2.1 FoV-Based Streaming

Duanmu et al. established a two-tier framework to intensify the utilization of bandwidth for 360° video streaming [16]. Skupin et al. propose an optimal way of streaming based on the FoV [17]. Sun et al. propose a two-tier solution to deliver the entire 360° span video at a truncated quality base tier and a higher quality enhancement tier [18]. Jiang et al. recommended Plato for viewport adaptive streaming using reinforcement learning [19].

Qian et al. introduce a cellular-friendly streaming methodology which conveys only 360° video viewport created on the prediction of head actions [20]. The 360° video stream has greater bandwidth requirements and needs quicker responsiveness to viewers' inputs [21]. In this aspect, Zhou et al. perform an analysis of oculus 360° video streaming. Among the future and past viewpoints, in order to capture the long-term dependent and nonlinear relation Yang et al. presented a single viewpoint prediction model built on CNN [22]. Corbillon et al. give a viewpoint adaptive approach that allows the streaming video to have a lower bit rate on comparison with the original video [23]. Table 4 gives the review on FoV-based streaming of 360° video.

Table 3 Summary on compression of 360° video

Author	Technique	Highlights	Challenges
Le et al. [10]	Transcoding and encryption methodology	Security is enhanced	–
Lin et al. [6]	360° certain coding tools	Highly effective in representing 360° videos	–
Mazumdar et al. [11]	Integrated perceptual compression approach	Maintaining the perceptual quality	Cost of compression is high
Xiu et al. [12]	Compression based on video coding methodology	Gains coding efficiency	Significant decoding complexity is increased
Wang et al. [7]	SCTMM with JEM scheme	SCTMM takes less time for decoding	Cost of SCTMM encoding is high
Zhang et al. [13]	Optimized compression algorithm	Improves efficiency	CU partition algorithm can be underestimated
Choi et al. [14]	High-quality video compression technique	Provides ordered representation of coding in a flexible manner	–
Lin et al. [15]	Containing various PEAs in large-scale database	Motivates perceptual video encoding mechanisms	–
Le et al. [10]	Real-time transcoding system	The two 4 K sessions and six 1080p sessions are optimized	–

2.2.2 Tile-Based Streaming

Sanchez et al. illustrate the streaming established by means of tile tactics followed in the Moving Picture Expert Group OMAF requirement [24]. Xie et al. presented a compatible streaming model for probabilistic tiles referred as 360ProbDASH [25]. Graf et al. propose adaptive tile-based streaming over HTTP to present the solution for the problems faced in video delivery infrastructures [26].

As the complexity of the 360° video increases with the essential to accomplish bitrate adaptation for a varying network [27], Le Feuvre and Concolato recommended MPEG DASH (Dynamic Adaptive-Streaming over HTTP) standard to designate by what means spatial accessing can be attained. Kammachi-Sreedhar and Curcio described an optimal way of streaming technology [28].

Nguyen et al. suggest a flexible method for tiling-based viewpoint streaming [29]. Due to the latency in the network, 360° video streaming is a difficult task [30]. Hence, Mahzari et al. recommended a tile-based caching policy. In real life using cellular networks, tiled video develops a probable solution for violently minimizing the essential bandwidth for 360° video transmission [31]. As a result, Lo et al. give

Table 4 Summary on FoV-based streaming of 360° video

Author	Technique	Highlights	Challenges
Duanmu et al. [16]	Two-tier mechanism	Achieves 25% gain	Bandwidth analysis is not focussed
Skupin et al. [17]	FoV-dependent optimal streaming	Enables ranking the FoV quality	Concurrent decoding for particular FoV is not done
Sun et al. [18]	Two-tier-based streaming scheme	Predicts FoV with high accuracy	–
Jiang et al. [19]	Viewport adaptive streaming	Qualities are adjusted based on FoV	–
Qian et al. [20]	Cellular-friendly 360° video delivery strategy	The client only yields the portions that are observable	To assess for a huge scale FoV, user study is challenging
Zhou et al. [21]	Analysis of Oculus 360° video streaming	Provides improved FoV visual quality	Improved performance is not analyzed
Yang et al. [22]	Single viewpoint prediction model	Enhances accuracy of FoV prediction	Spherical CNN is not explored in depth
Corbillon et al. [23]	Viewport-based adaptation algorithm	Viewport-dependent QER selection is done precisely	To predict the head movement is tedious

the performance over a cellular network of tile-based streaming. For high-quality streaming, there is a limitation of power consumption and bandwidth effectiveness [32]. Hence, Son et al. offer a tiling-based streaming approach. Summary on tile-based streaming of 360° video is shown in Table 5.

2.3 *Post-production of 360° Video*

At the user end, post-processing of the stored or streamed content is done. It provides consumer ease in comprehension, seamless visualization, and user experience. Several methods for post-production have been discussed under the following subsections.

2.3.1 Visualization

On live broadcasting, the broadcaster may not be aware of the user's FoV [33]. In this aspect, Takada et al. propose a visualization method based on users' Points of View (PoV) making use of a spherical heat map allowing the broadcaster to grip users' FoV easily and exchange information with users evenly. Azevedo et al. alter the 360° video signals for better visualization in immersive applications [1]. Existing

Table 5 Summary on tile-based streaming of 360° video

Author	Technique	Highlights	Challenges
Sanchez et al. [24]	Tile-based streaming approach	Observed fidelity was reduced	End-to End delay is critical
Xie et al. [25]	Probabilistic scheme for tiles	Provides contiguous playback	Saliency model needs to be precise
Graf et al. [26]	Adaptive tile-based streaming over HTTP	Evaluation of streaming is performed	–
Le Feuvre and Concolato [27]	MPEG DASH standard	Highly Interactive spatial navigation is achieved	Collaborative tile selection is not concentrated
Kammachi Sreedhar and Curcio [28]	Streaming technology based on adaptive bit rate	Multiple tiles synchronization	–
Nguyen et al. [29]	Tile-based viewpoint streaming	Tiles quality is improved	Optimality is not checked
Mahzari et al. [30]	Tile-based caching policy	Performance of cache is better	–
Lo et al. [31]	Tile-based streaming	Gives improved awareness	Transfer time depends on tile size
Son et al. [32]	Tile-based streaming	Transmit tiles autonomously	Lesser efficient

Table 6 Summary on visualization of 360° video

Author	Technique	Highlights	Challenges
Takada et al. [33]	Users PoV visualization method	Improves communication accuracy between users and senders	Functional improvement is required
Azevedo et al. [1]	Investigation on visualization	Proper psycho-visual examine on immersive applications is achieved	–

techniques, highlights, and challenges of visualization in 360° video are summarized in Table 6.

2.3.2 Viewport Prediction

User head movements result in user interaction and modifications in the spatial parts of the video allowing them to view only essential portions in the video for a specified time [9]. To achieve this, Aksu et al. offered a novel adaptable framework for the

Table 7 Summary on viewport prediction of 360° video

Author	Technique	Highlights	Challenges
Aksu et al. [9]	Viewport prediction with multicast live streaming scheme	Quality of viewport is maximized	–
Heyse et al. [34]	CB-based learning approach	Enhances Viewport quality	Further Enhancement is required
Jiang et al. [19]	Long short-term memory model	Future PoV is predicted early	–
Hu et al. [35]	Automated 360° piloting	Gives supreme performance	–
Sanchez et al. [24]	Viewpoint-dependent approach	Proves better visual resolutions	Significant gains can still be achieved
Li et al. [36]	Two groups prediction model	SD and mean of future PoV are predicted	–

prediction of the viewport. Heyse et al. offered an approach for contextual bandit based on reinforcement learning [34]. The tiles which map the field of view, provided with high resolution by using viewpoint adaptive streaming, was proposed by Jiang et al. [19]. Hu et al. recommended a mechanism of agent-based deep learning "deep 360° pilot" for viewers to pilot the 360° sports video spontaneously and develops an agent-specified domain to have a clear definition about the objects in the video [35].

To analyze visual quality at the viewport based on end-to-end delay, a viewpoint-dependent scheme was proposed by Sanchez et al. with the gain of 46% when compared with viewpoint-independent scheme [24]. Foreseeing the future PoV in a long time horizon can help in saving bandwidth incomes for on-request streaming of a video in which pausing of the video is diminished with noteworthy bandwidth variations in network [36]. To support this Li et al. introduced a two clusters point of view prediction models. Table 7 summarizes the viewport prediction of 360° video.

2.3.3 Designing Interface

Pavel et al. presented a technique based on the interactive orientation of shots enabling users to view all the significant content in the film [37]. Poblete et al. proposed a scalable appeal of design on crowdsourced technique [38]. Tang and Fakourfar supported collaborative perspective and interaction through proper awareness on gaze and technique on gesture for 360° videos [39]. The designing interfaces of 360° video are completely reviewed in Table 8.

Table 8 Summary on designing interface of 360° video

Author	Technique	Highlights	Challenges
Pavel et al. [37]	Interactive orientation of shots	Allows to choose significant points	Fatigue effect is not investigated
Poblete et al. [38]	Scalable appeal on crowdsource	Examines multiple fields of view	Lack of stitching multiple videos
Tang and Fakourfar [39]	Demonstrates the current view of interfaces	Powerful mechanism to be familiar with one another's view	Challenges faced are awareness of gaze and displaying in HMD

Table 9 Summary on user experience of 360° video

Author	Technique	Highlights	Challenges
Broeck et al. [40]	Numerous interaction methodologies	Highly ranking visualized experience	Results in motion sickness
Lin et al. [41]	Two focus guidance mechanism	Focus ease is improved	Multiple targets are not intensified
Nasrabadi et al. [42]	Taxonomy on 360° videos	Analysis with the varied clusters of users	Consequence of objects in motion is not studied in depth

Table 10 Summary on cybersickness of 360° video

Author	Technique	Highlights	Challenges
Bala et al. [43]	Study on existing methodologies	Guards from visually induced motion	Statistical implication is not encountered

2.3.4 User Experience

Broeck et al. proposed a numerous interaction methodology [40]. One task of looking at 360° videos is endlessly focusing and refocusing intentional targets [41]. To overcome this, Lin et al. addressed an approach on two focus guidance such as Automatic Piloting (directly taking audiences to the goal) and Visual Supervision (representing track of the goal). Nasrabadi et al. proposed taxonomy on 360° videos and classified them based on the motion of the camera and object [42]. Existing 360° video user experience techniques, highlights, and challenges are reviewed in Table 9.

2.3.5 Cybersickness

Bala et al. proposed an investigational study toward comparing and joining numerous available methodologies in 360° video to minimize cybersickness [43]. Cybersickness of 360° video is summarized in Table 10.

Table 11 Summary on summarization of 360° video

Author	Technique	Highlights	Challenges
Sung et al. [44]	Prototype based on memory network	Discourses narrative time-related summarization	–

Table 12 Summary on subtitle of 360° video

Author	Technique	Highlights	Challenges
Brown et al. [45]	Behaviors of subtitle	Answers each behavior usage	Diverse styles are not experimented

2.3.6 Summarization

For a long 360° videos, Sung et al. addressed the issue of story-based time-oriented summarization [44]. An innovative prototype based on memory network (Past Future Memory Network) was proposed. Available techniques, highlights, and challenges about summarization of 360° video are listed in Table 11.

2.3.7 Subtitle

Brown et al. designate behaviors of four subtitle (120-degree, static-follow, lag-follow, appear) in order to accomplish user testing in 360° video experience [45]. A detailed review of the subtitle of 360° video is illustrated in Table 12.

2.4 Quality Evaluation of 360° Video

This section gives the literature review on assessing the quality of the user-generated 360° videos. Some of the existing works have been listed in the following subsections.

2.4.1 Standardization

Wien et al. addressed the current status of standardization on focus with scientific aspects associated with the video [46]. Hannuksela et al. give an outline of the foremost edition of the standards in OMAF [47]. Skupin et al. presented the details regarding up-to-date status of precise efforts available in standardization [17]. Azevedo et al. offered some standardization techniques [1]. Domanski et al. proposed different kinds of visual media that are highly immersive [48]. Table 13 describes the standardization of 360° video techniques, highlights, and challenges.

Table 13 Summary on standardization of 360° video

Author	Technique	Highlights	Challenges
Wien et al. [46]	Outlines standardization efforts	Provides standards for coding and transmission	Standards for 6DoF remain unexplored
Hannuksela et al. [47]	Summary of the initial issue in OMAF	Embraces the representation of video setups, OMAF video profiles	Includes no abundant facts on image, audio, and text
Skupin et al. [17]	Describes the standardization status	Reports on the current status	–
Azevedo et al. [1]	Investigation of the most popular visual distortions	Acts as the basis for standardization	–
Domanski et al. [48]	Immersive standardization of visual media	Identify the immersive levels that are attained	Standardizing free perspectives are absent

Table 14 Summary on stabilization of 360° video

Author	Technique	Highlights	Challenges
Kopf [49]	Hybrid 2D-3D procedure	Gives improved smoothness	–
Tang et al. [50]	Combined stabilization approach	Demonstrates the stabilization of observing experience	–

2.4.2 Stabilization

Kopf offers a hybrid 2D-3D procedure for 360° video stabilizing by means of a deformed rotationally moving model [49]. Tang et al. introduce an approach for combined stabilization with the direction of 360° videos [50]. It includes a precisely designed new motion determination technique for 360° videos. Stabilization of 360° video is summarized in Table 14.

2.4.3 Assessment

Huang et al. support evaluation of video quality and propose a visual attention model for latitude-based 360-degree videos [51]. Hanhart et al. aim at the quality evaluation scheme recognized by JVET of ITU-T VCEG and ISO/IEC MPEG [52]. Zakharchenko et al. discussed the immersion media delivery format and quality assessment process [53]. Tran et al. investigated the quality benchmark of both subjective and objective for 360° videos [54]. For 360° video communication, Tran et

Table 15 Summary on assessment of 360° Video

Author	Technique	Highlights	Challenges
Huang et al. [51]	Visual attention model	Exploits the mean attention	Quality estimation in a large database is not investigated
Hanhart et al. [52]	Outlines the quality assessment framework	Gives an overview on quality assessment	–
Zakharchenko et al. [53]	Position-invariant quality metrics	Accurate and reliable	–
Tran et al. [54]	Quality benchmark	Assess the perceived quality	–
Tran et al. [55]	Study on the quality relationship	To detect suitable objective quality	Quality assessment on adaptive-based transmission is not focussed
Xie et al. [25]	QoE-based adaptation system	Variance of quality is minimized	–
Jiang et al. [19]	QoE metrics	Improvement on QoE	–
Corbillon et al. [23]	Multiple QER-based representation of a clip	Measurement of QoE only at the extracted PoV	–

al. aid to recognize suitable objective quality benchmark [55]. Xie et al. presented a QoE-based optimization framework [25]. Jiang et al. suggested Plato that outperforms existing strategy in numerous QoE metrics [19]. Corbillon et al. recommended an interactive high-quality mechanism for QoE measurement in Head-Mount-Device audience with small supervision [23]. Table 15 gives the quality assessment of 360° video.

3 Observation

The following are the observations made through this study:

- The 360° video is gaining interest among the consumers due to its simplicity.
- During projection there may be a chance of occurring visual artifacts which are also termed as distortion. Hence, extra caution has to be taken during the process of projection to learn the contents of the clip fruitfully.
- Once the 360° video is projected they undergo coding in order to have efficient storage and transmission where the 360° videos are compressed by preserving the quality of the video.
- As the viewpoint increases, the concept of streaming becomes difficult. Hence, an efficient approach of streaming the 360° video has to be done with better visual qualities.

- The high immersive nature of the video should not lead to motion sickness.
- 360° video can be delivered to the user optimally by summing up all the significant informations that are available in the clip.
- In order to have a clear understanding of the information available in the 360° clip, the video can be streamed with closed caption (i.e., text form).
- On the aspect of maximizing the smoothness of visual quality, the video can be stabilized.
- At the user end, the quality of the video can be checked by using the quality metrics.

4 Conclusion

360° video can offer an immersive experience for the users. As the FoV in 360° video increases in comparison with the standard videos, they encompass a huge amount of information. Due to the high resolution, 360° video processing, transmission, and displaying have to be done efficiently. This article presents the various techniques, highlights, and challenges involved in processing, transmission, and displaying the 360° video. At the viewer end, the decoded video has to be checked for its standardization, stabilization, and the quality of experience, to analyze the video for high standards, increased immersion, and improved QoE, respectively. The various techniques involved in the mechanism of standardization, stabilization, and QoE are listed in this survey with its highlights and challenges. The overall challenges faced in 360° videos are the high rate of compression and improvement in the quality-based viewport prediction.

On the aspect of future trends, the 360° videos are growing at a faster pace. In the near future, this technology will experience a huge leap. The major role of 360° video is storytelling with an immersive environment. Further improvement in terms of the cost may be possible in the coming years to give users an immersive experience. Faster improvement in the 360° technology and inexpensiveness of the equipment makes the 360° video to spread swiftly across many industries in the near future. In the upcoming years, for high-end performance, the 360° technology will provide a high level of video capturing with a High Dynamic Range (HDR).

References

1. R.G. de A. Azevedo, N. Birkbeck, F. De Simone, I. Janatra, B. Adsumilli, P. Frossard, Visual distortions in 360-degree videos, arXiv:1901.01848 (2019)
2. Y.-U. Yoon, D.-H. Park, J.-G. Kim, A method of padding inactive region for sphere segmented projection of 360° video, in *2018 International Workshop on Advanced Image Technology (IWAIT)*. IEEE (2018), pp. 1–3

3. K.-C. Huang, P.-Y. Chien, C.-A. Chien, H.-C. Chang, J.-I. Guo, A 360-degree panoramic video system design. Technical Papers of 2014 International Symposium on VLSI Design, Automation and Test, IEEE (2014), pp. 1–4
4. P. Hanhart, X. Xiu, Y. He, Y. Ye, 360-degree video coding based on projection format adaptation and spherical neighboring relationship, in *IEEE Journal on Emerging and Selected Topics in Circuits and Systems*, IEEE (2018)
5. Y.-C. Su, K. Grauman, Learning spherical convolution for fast features from 360 imagery, in *Advances in Neural Information Processing Systems* (2017), pp. 529–539
6. J.-L. Lin, Y.-H. Lee, C.-H. Shih, S.-Y Lin, H.-C. Lin, S.-K. Chang, P. Wang, L. Lin C.-C. Ju, Efficient Projection and coding tools for 360° Video. IEEE J. Emerging Selected Topics Circuits Syst. IEEE (2019)
7. Y. Wang, D. Liu, S. Ma, F. Wu, W. Gao, Spherical coordinates transform- based motion model for panoramic video coding. IEEE J. Emerg. Selected Topics Circuits Syst. IEEE (2019)
8. B. Vishwanath, T. Nanjundaswamy, K. Rose, Rotational motion model for temporal prediction in 360 video coding, in *2017 IEEE 19th International Workshop on Multimedia Signal Processing (MMSP)*. IEEE (2017), pp. 1–6
9. R. Aksu, J. Chakareski, V. Swaminathan, Viewport-driven rate-distortion optimized scalable live 360° video network multicast, in *2018 IEEE International Conference on Multimedia & Expo Workshops (ICMEW)*. IEEE (2018), pp. 1–6
10. T. Thanh Le, J.B. Jeong, E.-S. Ryu, Efficient transcoding and encryption for live 360 CCTV system. Applied Sciences, 9, 4, 760, Multidisciplinary Digital Publishing Institute (2019)
11. A. Mazumdar, B. Haynes, M. Balazinska, L. Ceze, A. Cheung, M. Oskin, Vignette: perceptual compression for video storage and processing systems, arXiv:1902.01372 (2019)
12. X. Xiu, Y. He, Y. Ye, R. Vanam, H. Philippe, T. Lu, E. Pu, P. Yin, W. Husak, T. Tao., Improved video coding techniques for next generation video coding standard, in *2019 Data Compression Conference (DCC)*. IEEE (2019), pp. 290–299
13. M. Zhang, J. Zhang, Z. Liu, C. An, An efficient coding algorithm for 360-degree video based on improved adaptive qp compensation and early cu partition termination, in *Multimedia Tools and Applications*, 78, 1. Springer (2019), pp. 1081–1101
14. K. Choi, J. Chen, A. Tamse, H. Yang, M.W. Park, S. Ikonin, W. Choi, S. Esenlik, New video codec for high- quality video service and emerging applications, in *2019 Data Compression Conference (DCC)*. IEEE (2019), pp. 310–319
15. L. Lin, S. Yu, T. Zhao, Z. Wang, others, PEA265: Perceptual assessment of video compression artifacts, arXiv:1903.00473 (2019)
16. F. Duanmu, E. Kurdoglu, Y. Liu, Y. Wang, View direction and bandwidth adaptive 360 degree video streaming using a two-tier system, in *2017 IEEE International Symposium on Circuits and Systems (ISCAS)*. IEEE (2017), pp. 1–4
17. R. Skupin, Y. Sanchez, Y.-K. Wang, M.M. Hannuksela, J. Boyce, M. Wien, Standardization status of 360 degree video coding and delivery, in *2017 IEEE Visual Communications and Image Processing (VCIP)*. IEEE (2017), pp. 1–4
18. L. Sun, F. Duanmu, Y. Liu, Y. Wang, Y. Ye, H. Shi, D. Dai, A two- tier system for on- demand streaming of 360 degree video over dynamic networks. IEEE J. Emerg. Selected Topics Circuits Syst. IEEE (2019)
19. X. Jiang, Y.-H. Chiang, Y. Zhao, Y. Ji, Plato: learning- based adaptive streaming of 360-degree videos, in *2018 IEEE 43rd Conference on Local Computer Networks (LCN)*. IEEE (2018), pp. 393–400
20. F. Qian, L. Ji, B. Han, V. Gopalakrishnan, Optimizing 360 video delivery over cellular networks, in *Proceedings of the 5th Workshop on All Things Cellular: Operations, Applications and Challenges*. ACM (2016), pp. 1–6
21. C. Zhou, Z. Li, Y. Liu, A measurement study of oculus 360 degree video streaming, in *Proceedings of the 8th ACM on Multimedia Systems Conference*. ACM (2017), pp. 27–37
22. Q. Yang, J. Zou, K. Tang, C. Li, H. Xiong, Single and sequential viewports prediction for 360-degree video streaming, in *2019 IEEE International Symposium on Circuits and Systems (ISCAS)*. IEEE (2019), pp. 1–5

23. X. Corbillon, G. Simon, A. Devlic, J. Chakareski, Viewport- adaptive navigable 360-degree video delivery, in *2017 IEEE International Conference on Communications (ICC)*. IEEE (2017), pp. 1–7
24. Y. Sanchez, G.S. Bhullar, R. Skupin, C. Hellge, T. Schierl, Delay Impact on MPEG OMAF's tile-based viewport- dependent 360° video streaming. IEEE J. Emerg. Selected Topics Circuits Syst. IEEE (2019)
25. L. Xie, Z. Xu, Y. Ban, X. Zhang, Z. Guo, 360probdash: improving qoe of 360 video streaming using tile-based Http adaptive streaming, in *Proceedings of the 25th ACM international conference on Multimedia*. ACM (2017), pp. 315–323
26. M. Graf, C. Timmerer, C. Mueller, Towards bandwidth efficient adaptive streaming of omnidirectional video over Http: design, implementation, and evaluation, in *Proceedings of the 8th ACM on Multimedia Systems Conference*. ACM (2017), pp. 261–271
27. J. Le Feuvre, C. Concolato, Tiled- based adaptive streaming using MPEG-DASH, in *Proceedings of the 7th International Conference on Multimedia Systems*. ACM (2016), p. 41
28. K. Kammachi-Sreedhar, I.D.D. Curcio, Omnidirectional video delivery with decoder instance reduction, in *Internet Technology Letters*, vol. 2, 1, e79, Wiley Online Library (2019)
29. D.V. Nguyen, H.T.T. Tran, T.C. Thang, Adaptive tiling selection for viewport adaptive streaming of 360-degree video, in *IEICE Transactions on Information and Systems*, vol. 102, 1. The Institute of Electronics, Information and Communication Engineers (2019), pp. 48–51
30. A. Mahzari, A. Taghavi Nasrabadi, A. Samiei, R. Prakash, Fov- aware edge caching for adaptive 360 video streaming, in *2018 ACM Multimedia Conference on Multimedia Conference*. ACM (2018), pp. 173–181
31. W.-C. Lo, C.-L. Fan, S.-C. Yen, C.-H. Hsu, Performance measurements of 360° video streaming to head-mounted displays over live 4G cellular networks, in *2017 19th Asia-Pacific Network Operations and Management Symposium (APNOMS)*. IEEE (2017), pp. 205–210
32. J. Son, D. Jang, E.-S. Ryu, Implementing 360 video tiled streaming system, in *Proceedings of the 9th ACM Multimedia Systems Conference*. ACM (2018), pp. 521–524
33. M. Takada, D. Nishioka, Y. Saito, Proposal of a spherical heat map in 360-degree internet live broadcasting using viewers' POV, in *2019 IEEE International Conference on Pervasive Computing and Communications Workshops (PerCom Workshops)*. IEEE (2019), pp. 596–600
34. J. Heyse, M.T. Vega, F. De Backere, F. De Turck, Contextual bandit learning-based viewport prediction for 360 video, in *IEEE Virtual Reality (VR)* (2019)
35. H.-N. Hu, Y.-C. Lin, M.-Y. Liu, H.-T. Cheng, Y.-J. Chang, M. Sun, Deep 360 pilot: learning a deep agent for piloting through 360 sports videos, in *2017 IEEE Conference on Computer Vision and Pattern Recognition (CVPR)*. IEEE (2017), pp. 1396–1405
36. C. Li, W. Zhang, Y. Liu, Y. Wang, Very long term field of view prediction for 360-degree video streaming, in *2019 IEEE Conference on Multimedia Information Processing and Retrieval (MIPR)*. IEEE (2019), pp. 297–302
37. A. Pavel, B. Hartmann, M. Agrawala, Shot orientation controls for interactive cinematography with 360 video, in *Proceedings of the 30th Annual ACM Symposium on User Interface Software and Technology*. ACM (2017), pp. 289–297
38. B.M. Poblete, E.C. Mendoza, J.P. De Castro, J.A. Deja, G. Nodalo, A research through design (Rtd) approach in the design of a 360-video platform interface, in *Proceedings of the 5th International ACM In-Cooperation HCI and UX Conference*. ACM (2019), pp. 166–171
39. A. Tang, O. Fakourfar, Watching 360 videos together, in *Proceedings of the 2017 CHI Conference on Human Factors in Computing Systems*. ACM (2017), pp. 4501–4506
40. M.V. den Broeck, F. Kawsar, J. Schöning, It's all around you: exploring 360 video viewing experiences on mobile devices, in *Proceedings of the 25th ACM international conference on Multimedia*. ACM (2017), pp. 762–768
41. Y.C. Lin, Y.-J. Chang, H.-N. Hu, H.-T. Cheng, C.-W. Huang, M. Sun, Tell me where to look: investigating ways for assisting focus in 360 video, in *Proceedings of the 2017 CHI Conference on Human Factors in Computing Systems*. ACM (2017), pp. 2535–2545
42. A.T. Nasrabadi, A. Samiei, A. Mahzari, R.P. McMahan, R. Prakash, M.C.Q. Farias, M.M. Carvalho, A taxonomy and dataset for 360° videos, in *Proceedings of the 10th ACM Multimedia Systems Conference*. ACM (2019), pp. 273–278

43. P. Bala, D. Dionísio, V. Nisi, N. Nunes, Visually induced motion sickness in 360° videos: comparing and combining visual optimization techniques, in *2018 IEEE International Symposium on Mixed and Augmented Reality Adjunct (ISMAR-Adjunct)*. IEEE (2018), pp. 244–249
44. S. Lee, J. Sung, Y. Yu, G. Kim, A memory network approach for story-based temporal summarization of 360 videos, in *Proceedings of the IEEE Conference on Computer Vision and Pattern Recognition* (2018), pp. 1410–1419
45. A. Brown, J. Turner, J. Patterson, A. Schmitz, M. Armstrong, M. Glancy, Subtitles in 360-degree video, in *Adjunct Publication of the 2017 ACM International Conference on Interactive Experiences for TV and Online Video*. ACM (2017), pp. 3–8
46. M. Wien, J.M. Boyce, T. Stockhammer, W.H. Peng, Standardization status of immersive video coding. IEEE J. Emerg. Selected Topics Circuits Syst., IEEE (2019)
47. M.M. Hannuksela, Y.K. Wang, A. Hourunranta, An overview of the OMAF standard for 360° video, in *2019 Data Compression Conference (DCC)*. IEEE (2019), pp. 418–427
48. M. Domański, O. Stankiewicz, K. Wegner, T. Grajek, Immersive visual media- mpeg-i: 360 video, virtual navigation and beyond, in *2017 International Conference on Systems, Signals and Image Processing (IWSSIP)*. IEEE (2017), pp. 1–9
49. J. Kopf, 360 video stabilization. ACM Trans. Graph. (TOG) **35**(6), 195 (2016), ACM
50. C. Tang, O. Wang, F. Liu, P. Tan, Joint stabilization and direction of 360° videos. ACM Trans. Graph. (TOG) **38**(2), 18 (2019), ACM
51. H. Huang, Y. Xu, J. Chen, S. Song, T. Zhao, Latitude-based visual attention in 360-degree video display, in *Pacific Rim Conference on Multimedia*. Springer (2018), pp. 282–290
52. P. Hanhart, Y. He, Y. Ye, J. Boyce, Z. Deng, L. Xu, 360-degree video quality evaluation, in *2018 Picture Coding Symposium (PCS)*. IEEE (2018), pp. 328–332
53. V. Zakharchenko, K.P. Choi, J.H. Park, Quality metric for spherical panoramic video, optics and photonics for information processing X, 9970, 99700C. International Society for Optics and Photonics (2016)
54. H.T.T. Tran, C.T. Pham, N.P. Ngoc, C.M. Bui, M.H. Pham, T.C. Thang, An evaluation of quality metrics for 360 videos, in 2017 *Ninth International Conference on Ubiquitous and Future Networks (ICUFN)*. IEEE (2017), pp. 7–11
55. H.T.T. Tran, C.T. Pham, N.P. Ngoc, A.T. Pham, T.C. Thang, A study on quality metrics for 360 video communications, in *IEICE TRANSACTIONS on Information and Systems*, vol. 101, 1. The Institute of Electronics, Information and Communication Engineers (2018), pp. 28–36

A Study of Scrambled Noisy Quantum Image Formation with Geometric Transformation and Its Denoising Using QWT

S. Chakraborty, S. H. Shaikh, A. Chakrabarti, and R. Ghosh

Abstract Quantum image processing reduces the gap between quantum computing and image processing fields. The principles of quantum computing explore the image processing in various ways of handling (capture, manipulate, extract) images of different formats and for different purposes. In this paper, an image denoising scheme based on quantum wavelet transform (QWT) is proposed. Initially, a quantum noisy image is formed with the help of geometric transformation operation and embedded into the wavelet coefficients of the quantum original grayscale image. As a result, it will affect the visual quality of the original quantum image. Then the quantum Daubechies wavelet kernel of fourth order is used to extract wavelet coefficients from the resultant image. Then we consider a Quantum Oracle that implements a suitable thresholding function to decompose the wavelet coefficients into a greater effect applicable for the original image wavelet coefficients and lower effect for the noisy image wavelet coefficients. However, original image wavelet coefficients are greater than the noisy wavelet coefficients. A detail computational time and storage complexity analysis is given and compared with some state-of-the-art denoising techniques at the end of this paper.

S. Chakraborty (✉) · A. Chakrabarti · R. Ghosh
A.K.C.S.I.T., Calcutta University, Kolkata, India
e-mail: schakraborty770@gmail.com

A. Chakrabarti
e-mail: achakra12@yahoo.com

R. Ghosh
e-mail: rghosh47@gmail.com

S. H. Shaikh
CSE, BML Munjal University, Gurugram, India
e-mail: soharab.hossain@gmail.com

A. Nanda and N. Chaurasia (eds.), *High Performance Vision Intelligence*,
Studies in Computational Intelligence 913,
https://doi.org/10.1007/978-981-15-6844-2_10

1 Introduction

The most important way of pictorial representations of information is an image. It represents information with various styles that have a huge number of applications on the fields of medical science, biomedicine, meteorology, telecommunication, satellite communication, etc. The immense computing power offered by the realization of a quantum computer has led to an increasing interest in the field of image processing. This computational power draws toward the use of three major principles coming from quantum physics: quantum entanglement of states, quantum interference, and quantum parallelism [9]. From a comparison point of view, a quantum computer is superior than the classical computer in terms of efficiency of solving problems quicker such as Deutsch and Jozsa's algorithm introduced a checking constraint to decide whether a function is even or balanced, Shor's factoring large integers, Grover's database searching algorithm [9]. Because of their high efficiency in terms of computational time complexity, these two quantum algorithms are used extensively. Table 1 shows a short comparison between some popular classical algorithms and its quantum counterparts in terms of computational time complexity.

To deal with quantum images, the primary task is to convert the classical images into quantum counterpart. However, several popular methods have been developed that are mainly focused on the storage models of the quantum image, such as qubit lattice, entangled image, multichannel representation of quantum images (MCQI), flexible representation of quantum images (FRQI), and novel enhanced quantum representation of digital image model (NEQR) [3]. With the study of quantum image representation, researchers begin to research quantum image security, compression, filtering, etc. [10, 11]. Images collected in raw format are equipped with various types of noises. These noises are responsible for deteriorating the quality of images and create problems in diagnosis and interpretation processes. Any denoising technique focuses to remove these noisy components while preserving the important signal as much as possible [7, 16]. However, several works on image denoising have been done classically using Fourier transform and wavelet transform [6, 15, 16]. But no work has been done on image denoising in the quantum domain till now as there is a primary obstacle which is called unitary transformations for describing any quantum algorithms. The desired computation on a quantum computer must be unitary, linear, and reversible in nature. An extra "ancilla" qubit helps us to handle irreversibility. But the problem still exists in the quantum domain due to nonlinear transformations.

Fortunately, classical Fourier and wavelet transforms can be described with the help of unitary operators. This is the reason, quantum Fourier transform (QFT) and quantum wavelet transform (QWT) are extensively studied by the quantum computing community nowadays. However, QFT is also recognized as a pivotal application in many popular quantum algorithms. The wavelet transform is an extension of Fourier transform and it is also very useful in the context of classical computing.

To deal with the noise problems, few researchers are focusing on some filtering processes in quantum images [8, 10]. A quantum Fourier transform (QFT)-based quantum image filtering model is proposed and implemented in the paper [8, 10].

Table 1 Comparison between classical and quantum algorithms

	FFT	Wavelet transform	Walsh transform	Search algorithm	Prime factorization
Classical	O($n2^n$)	O(2^n)	O($n2^n$)	O(n)	$e^{O(n^{\frac{1}{3}} log^{\frac{2}{3}} n)}$
Quantum	O(n^2)	O(n^2)	O(n)	O($\sqrt{n}$)	O($n^2 lognloglogn$)

Although QFT provides an exponential speedup over its classical FFT, it is very hard to exploit because the set of Fourier coefficients are stored as probability amplitudes of quantum states and thus cannot be directly accessed. However, Fourier transform and inverse Fourier transform are applied as the beginning and ending steps of this algorithm [10]. The major drawback of this method is that the internal view of Quantum Oracle is unknown. But in our proposed technique, we have shown the detailed design of the internal quantum circuit of the proposed Quantum Oracle. In a paper [8], image filtering has been performed with the help of a filter mask association. However, quantum addition operation is used for this correlation instead of quantum multiplication in paper [8]. In a paper [11], quantum image denoising has been achieved using a quantum median filter in the spatial domain. In this paper, a median fialter has been made of four basic quantum modules (i.e., Cycle Shift, Comparator, and Swap).

Inspired from the above quantum filtering approaches, our main focus is to introduce a quantum image denoising technique using quantum wavelet transform in this chapter. We use the concept of Daubechies fourth-order wavelet kernel [1] along with the elementary quantum geometric transformation operation (for making scrambled quantum noisy image) [15] and quantum addition operation. Finally, we use a specialized Quantum Oracle along with quantum parallelism operation to filter the necessary quantum wavelet coefficients.

2 Background

2.1 Quantum Computing

Qubit is the measure as a basic unit of quantum computing circuit. Binary quantum system exists in a linear superposition of two basis states, labeled by $|0\rangle$ and $|1\rangle$. In binary information processing, $|0\rangle$ and $|1\rangle$ qubit states are not only used to store information but also the superposition of $|0\rangle$ and $|1\rangle$, like

$$|\psi\rangle = \alpha|0\rangle + \beta|1\rangle, \tag{1}$$

where α and β represent probability amplitudes of the basis states in a Hilbert space ($H = C^2$) and the probability of storing the information is higher to $|0\rangle$ basis state if

$|\alpha|^2$ is greater than $|\beta|^2$ and vice versa [9, 10]. The basic unit to represent information in binary quantum system is qubit. So the superposition state $|\psi\rangle$ can be represented as

$$|\psi\rangle = \sum_{j=0}^{2^n-1} \alpha_j |j\rangle, \tag{2}$$

where $|\alpha_j|^2$ represents the probability of getting value $|j\rangle$. A quantum register is a storage of qubits and a quantum gate is an application of unitary operator (U) over quantum register in state space H [10]. Quantum search algorithms also offer numerous algorithmic speedup than their classical counterpart using some extraordinary properties of quantum mechanics like superposition, interference, entanglement, etc. In our proposed work, we use quantum network along with quantum arithmetic and quantum thresholding operation to implement the quantum denoising scheme. However, all the arithmetic operations in quantum are unitary and reversible in nature [9, 10]. Some elementary quantum gates such as NOT gate, controlled-NOT gate, Toffoli gate, swap gate, phase change gate, and so on are described in the paper [5].

2.2 *Wavelet Transformation in Binary Quantum System*

In classical computation, wavelet transform plays a vital role in image denoising, compression, and processing [6, 12]. A fast and efficient quantum circuit for two popular wavelet transforms is first introduced in paper [1]. At the back end of Daubechies $(D^{(4)})$ wavelet, the quantum Fourier transform (QFT) circuit plays the key role. In our work, we are using fourth-order Daubechies wavelet kernel that has 2^n dimensions for image processing. The main aim of this technique is to factorize the transform matrix into elementary unitary matrices. The overall quantum representation of Daubechies $D^{(4)}_{2^n}$ fourth-order wavelet kernel and its corresponding quantum circuit is described below in Eq. 3 [1],

$$D^{(4)}_{2^n} = (I_{2^{n-1}} \otimes C_1) Q_{2^n} (I_{2^{n-1}} \otimes C'_0), \tag{3}$$

where C_1 and C'_0 are two unitary matrices that can be represented as

$$C'_0 = 2 \begin{bmatrix} C_3 & -C_2 \\ C_2 & C_3 \end{bmatrix}$$

and

$$C_1 = \frac{1}{2} \begin{bmatrix} \frac{C_0}{C_3} & 1 \\ 1 & \frac{C_1}{C_2} \end{bmatrix}$$

wherein, $C_0 = (3 + \sqrt{3})/4\sqrt{2}$, $C_1 = (3 - \sqrt{3})/4\sqrt{2}$, $C_2 = (1 - \sqrt{3})/4\sqrt{2}$, $C_3 = (1 + \sqrt{3})/4\sqrt{2}$, $C'_0 = N(C_0)$

$$N = \begin{bmatrix} 0 & 1 \\ 1 & 0 \end{bmatrix}$$

and Q_{2^n} is a downshift permutation matrix and it is given below:

$$Q_{2^n} = \begin{bmatrix} 0\,1 \\ 0\,0\,1 \\ 0\,0\,0\,1 \\ 0\,0\,0\,0\,1 \\ \cdot\;\cdot\;\cdot\;\cdot\;\cdot\;\cdot \\ \cdot\;\cdot\;\cdot\;\cdot\;\cdot\;\cdot \\ 0\,0\,.\,.\,.\,.\,0\,0\,0 \\ 1\,0\,.\,.\,.\,.\,0\,0\,0 \end{bmatrix}.$$

However, this permutation matrix can be factored as

$$Q_{2^n} = F_{2^n} P_{2^n} T_{2^n} P_{2^n} F^*_{2^n}, \tag{4}$$

where the classical Cooley–Tukey FFT factorization for a 2n-dimensional vector is represented by F_{2^n} in Eq. 4, $F^*_{2^n}$ is represented as the conjugate transpose of F_{2^n} [1], and P_{2^n} is a bit-reversal permutation matrix. If FFT (fast Fourier transform) is used as the first stage of the computation and to make reverse order input qubits we will take the help of the Cooley–Tukey FFT. P_{2^n} can be represented as

$$P_{2^n} : |a_{n-1}a_{n-2}......a_1a_0\rangle \rightarrow |a_0a_1........a_{n-2}a_{n-1}\rangle \tag{5}$$

and T_{2^n} is a diagonal matrix given as $T_{2^n} = diag[1, \omega_{2^n}, \omega^2_{2^n}, ...\omega^{2^n-1}_{2^n}]$, $\omega^2_{2^n} = e^{\frac{-2i\pi}{2^n}}$, $i = \sqrt{-1}$. However, the final product matrix $P_{2^n}T_{2^n}P_{2^n}$ can be factorized into

$$\begin{aligned} P_{2^n}T_{2^n}P_{2^n} = (I_{2^{n-1}} \otimes G(\omega^{2^{n-1}}_{2^n}))...(I_{2^{n-i}} \otimes G(\omega^{2^{n-i}}_{2^n}) \\ \otimes I_{2^{n-1}})...(G(\omega_{2^n}) \otimes I_{2^{n-1}}), \end{aligned} \tag{6}$$

wherein $G(\omega^k_{2^n}) = diag(1, \omega^k_{2^n})k = 0, 1, ..., n-1$, and

$$G(\omega^k_{2^n}) = \begin{bmatrix} 1 & 0 \\ 0 & \omega^k_{2^n} \end{bmatrix}.$$

Figure 1 shows a fourth-order Daubechies wavelet kernel circuit.

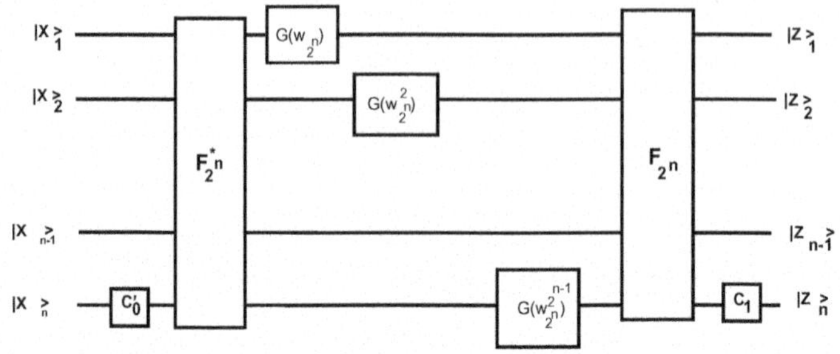

Fig. 1 Quantum circuit of Daubechies fourth-order wavelet kernel [1]

3 Proposed Methodology

In the case of the classical approach, the basic steps for doing threshold-based image denoising using wavelet transform are given as follows:

1. Add noisy (Gaussian or salt-pepper noise) image with the original image (geometrical transformed image in our case).
2. Perform a multilevel discrete wavelet decomposition on the corrupted image.
3. Identify a thresholding technique to select a suitable threshold range.
4. Perform hard or soft thresholding over the resultant wavelet coefficients.
5. Extract and reconstruct the original image by the help of applying inverse discrete wavelet transform.

However, in our work, we have presented a new quantum image denoising scheme using quantum wavelet transform (QWT). The entire denoising procedure is classified into some sub-procedures.

3.1 Generation of a Quantum Noisy Image and the Original Quantum Image

The proposed image denoising scheme uses the famous FRQI procedure [3, 4] to represent an original grayscale image and we would generate a noisy image by introducing a fixed pattern quantum phase change operation to replace a certain number of pixels in a quantum way. Because QWT and the subsequent denoising algorithms are easy to carry out in FRQI representation. Given an original image of size $2^n \times 2^n$,

$$|Q\rangle = \frac{1}{2^n} \sum_{t=0}^{2^{2n-1}} (\cos\theta_t|0\rangle + \sin\theta_t|1\rangle) \otimes |t\rangle \tag{7}$$

$$|Q\rangle = \sum_{t=0}^{2^{2n-1}} (|c_t\rangle) \otimes |t\rangle, \theta_t \in [0, \frac{\pi}{2}], t = 0, 1, 2, ..., 2^{2n-1} \tag{8}$$

$$\begin{aligned} |Q\rangle = \sum_{t=0}^{2^{2n-1}} (|c_t\rangle) \otimes |y_t\rangle|x_t\rangle, |y_t\rangle = |y_{n-1}y_{n-2}......y_0\rangle\ and \\ |x_t\rangle = |x_{n-1}x_{n-2}......x_0\rangle, \end{aligned} \tag{9}$$

where $|y_t\rangle|x_t\rangle\ belongs\ to\ the\ range\ of [0, 1], t = 0, 1, 2, ..., n$. Here, $|c_t\rangle$ represents color information and $|y_t\rangle|x_t\rangle$ represents position information of the pixels of a quantum image.

3.2 Geometric Transformation of FRQI Image to Build Its Noisy Version

However, geometric transformation (G_Q) of the quantum FRQI image sized $2^n \times 2^n$ can be represented as

$$G_Q(|Q\rangle) = \frac{1}{2^n} \sum_{t=0}^{2^{2n-1}} (|c_t\rangle) \otimes G(|t\rangle), \tag{10}$$

where $G(|t\rangle)$ for $t = 0, 1, 2, ..., 2^{2n-1}$ which perform information interchange geometrically based on the vertical and horizontal locations. The function G is applied on computational basis vectors as shown in Fig. 2. However, we can perform such geometric transformation on FRQI image using the concept of two-point swapping operations [15]. FRQI image can be swapped between two positions x and y and this can be represented as

$$S(|Q\rangle) = |PN\rangle = \frac{1}{2^n} \sum_{k=0}^{2^{2n-1}} (|c_k\rangle) \otimes S(|k\rangle), \tag{11}$$

$$\begin{aligned} where\ S(|k\rangle) = |k\rangle, k \neq x, y\ and\ S(|x\rangle) = |y\rangle, S(|y\rangle) = |x\rangle, \\ i.e., S = |x\rangle\langle y| + |y\rangle\langle x| + \sum^{k \neq x,y} |k\rangle\langle k|. \end{aligned} \tag{12}$$

However, we will describe the performance of S over the superposition of $|k\rangle$ for $k = 0, 1, 2, ..., 2^{2n-1}$, i.e., $|k\rangle = \sum_{k=0}^{2^{2n-1}} \alpha_k |k\rangle$. For sake of simplicity, if we consider a 2×2 FRQI image includes four points as in Fig. 3 which are represented by four encoded bits string as $|00\rangle, |10\rangle, |01\rangle, |11\rangle$. According to Fig. 3, $|10\rangle|11\rangle$ and

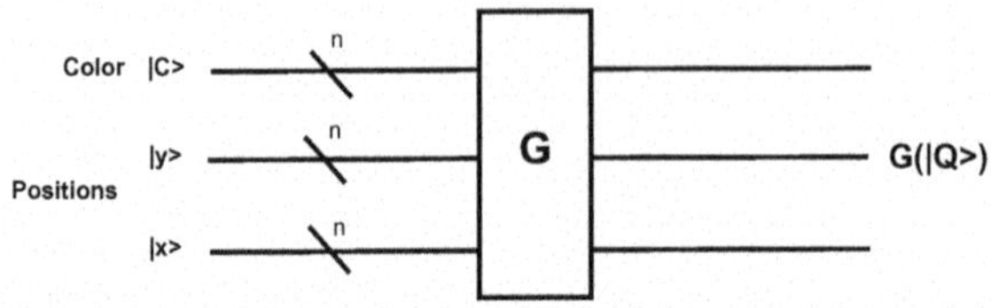

Fig. 2 A quantum image geometric transformation circuit

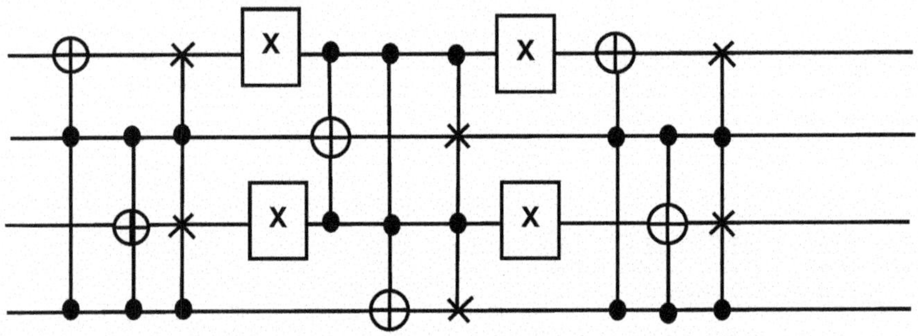

Fig. 3 A quantum circuit of two-point swapping operation on 2×2 image

	00	01	10	11
00				
01		120	89	
10		68	150	**200**
11				

	00	01	10	11
00		**200**		
01		120	89	
10		68	150	
11				

Fig. 4 Two-point swapping operation on an image matrix

$|10\rangle$, $|01\rangle$ points are swapped due to the impact of the two-point swapping operation. The quantum circuit of this two-point swap operation on a 2×2 image is shown in Fig. 4. This transformed image ($|GN\rangle$) will be considered as a noisy (scrambled) image of the corresponding FRQI image for the next operation.

3.3 Apply QWT on the Original Quantum Image

In this step, we execute QWT (Duabechies $D^{(4)}$ wavelet transform [1]) on the original quantum image $|Q\rangle$.

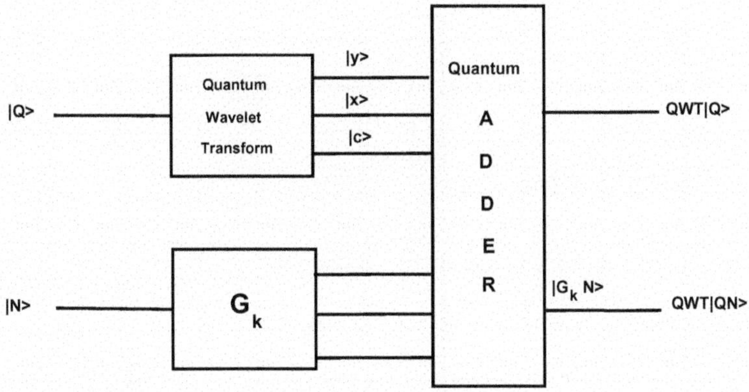

Fig. 5 A quantum noisy image embedding operation

$$
\begin{aligned}
QWT(|Q\rangle) &= QWT\left(\frac{1}{2^n}\sum_{t=0}^{2^{2n-1}}(\cos\theta_t|0\rangle + \sin\theta_t|1\rangle)\otimes|t\rangle\right) \\
&= \frac{1}{2^n}\sum_{t=0}^{2^{2n-1}} QWT(\cos\theta_t|0\rangle + \sin\theta_t|1\rangle\otimes|t\rangle\Big) \\
&= \sum_{t=0}^{2^{2n-1}}|wc_t\rangle\otimes|t>.
\end{aligned} \tag{13}
$$

3.4 Embedding Operation of the Noisy Image

The transformed quantum noisy image $|GN\rangle$ is embedded into the wavelet coefficients $QWT(|Q\rangle)$ and we obtain

$$
QWT(|QN\rangle) = QWT(|Q\rangle) + |G_k N\rangle) = \sum_{i=0}^{2^{2n-1}}|wmc_i\rangle\otimes|i>, \tag{14}
$$

where $QWT(|Q\rangle)$ represents the QWT qubits of the original image and $QWT(|QN\rangle)$ represents the qubits after embedding the noisy image. This embedding procedure is completely reversible. A basic quantum adder network is used to implement this embedding process is shown in Fig. 5. The original image and its corresponding noisy images are stored and encoded in two quantum registers initially. Then, the $DB^{(4)}$ wavelet kernel operation is carried out on the original quantum image and a series of G_k operations are also used in this embedding process. The basic quantum sum and carry network which help to build the circuit of quantum adder are shown in Fig. 6.

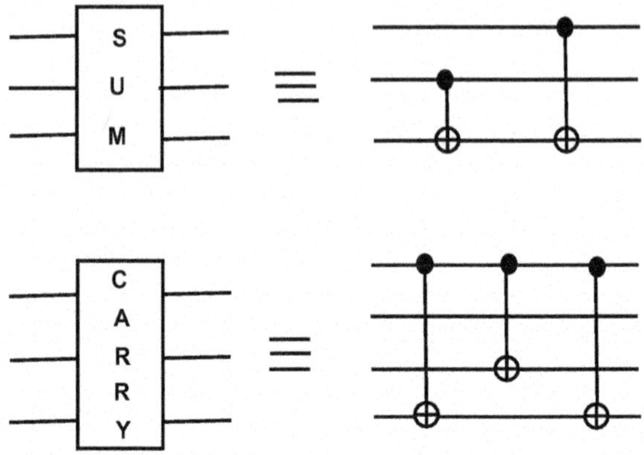

Fig. 6 A quantum adder network to represent basic sum and carry operations

3.5 *Thresholding Operation on the Resultant Image Using Quantum Oracle*

In order to achieve the thresholding operation in the frequency domain, we propose a Quantum Oracle provided as a black box [3, 13]. In this step, we have to apply a threshold-based operation to classify low- and high-coefficient components from the final product of the previous step $QWT(|QN\rangle)$ and the additional qubit $|0\rangle$ can be used to make the distinction between the high- or low-frequency images. In this procedure, we apply the thresholding function $f : (0, 1, 2, \ldots\ldots, 2^{m+2n} - 1) \rightarrow (0, 1)$ in the form of Oracle operator U_f on the state $QWT|QN\rangle \otimes |0\rangle$ (refer Eq. 15). The principle of quantum parallelism also takes an active role in this thresholding operation. The input state $|I_0>$ of our Quantum Oracle is represented by

$$\begin{aligned} |I_0\rangle &= |H_{cf}\rangle^{\otimes m}|L_{cf}\rangle^{\otimes m}|TH\rangle_m QWT|QN\rangle_{m+2n}|0\rangle \\ &= |1\rangle^{\otimes m}|0\rangle^{\otimes m}U_f(|TH\rangle_m QWT|QN\rangle_{m+2n}|0\rangle) \\ &= |1\rangle^{\otimes m}|0\rangle^{\otimes m}(QWT|QN^{Hf}\rangle|0\rangle + QWT|QN^{Lf}\rangle|1\rangle). \end{aligned} \tag{15}$$

The color register's state is set to $|1\rangle^{\otimes m}$ when the Oracle qubit is $|1\rangle$ (representing high-frequency (H_{cf}) components) and to $|0\rangle$ (representing low-frequency (L_{cf}) components) when the Oracle qubit is $|0\rangle$. This state change operation is done by cSWAP gate shown in the quantum circuit of Fig. 7. $|TH\rangle$ register holds the threshold value. The operation of U_{CMP} [20] can be described by

$$U_{CMP}|x\rangle|y\rangle|0\rangle^{\otimes p}|0\rangle|0\rangle = |x\rangle|y\rangle|\psi\rangle|u_1\rangle|u_2\rangle, \tag{16}$$

where $|x\rangle$ and $|y\rangle$ are p-qubit registers that hold the compared states, $p + 2$ ancillae and $|u_1\rangle$, $|u_2\rangle$ contain the comparison results.

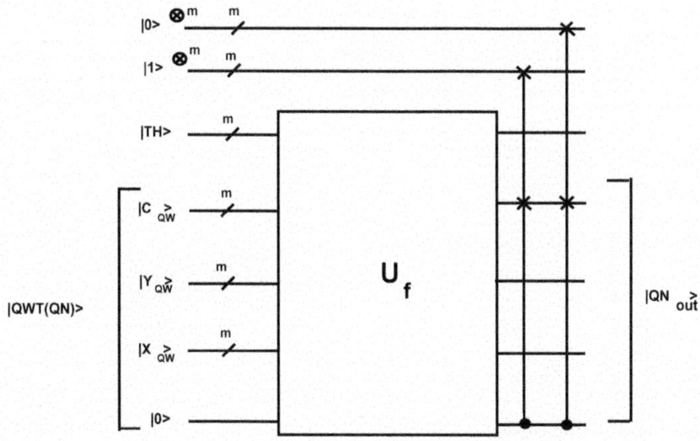

Fig. 7 A quantum network of thresholding wavelet coefficients of processed image

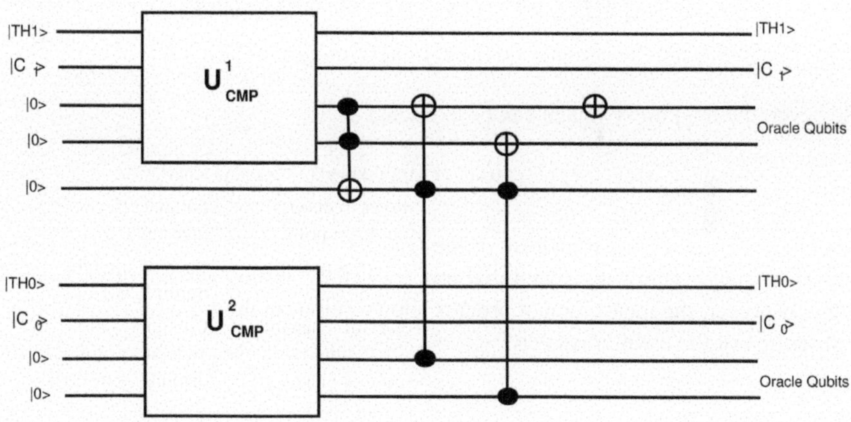

Fig. 8 A Quantum Oracle circuit using quantum comparator

$$
\begin{aligned}
|u_1\rangle &= |u_2\rangle = 0, if x = y \\
|u_1\rangle &= 1, |u_2\rangle = 0, if x > y. \\
|u_1\rangle &= 0, |u_2\rangle = 1, if x < y
\end{aligned}
\tag{17}
$$

The quantum circuit design of comparison operator and our Oracle operator are realized in Fig. 8. This Oracle operator U_f inverts the Oracle qubit if $c \geq TH$, where $TH = |TH_1\rangle|TH_0\rangle$ represents the threshold and $|C\rangle = |C_1\rangle|C_0\rangle$ represents the gray level. If the input image size is increased then the implementation cost of our Quantum Oracle is also increased linearly.

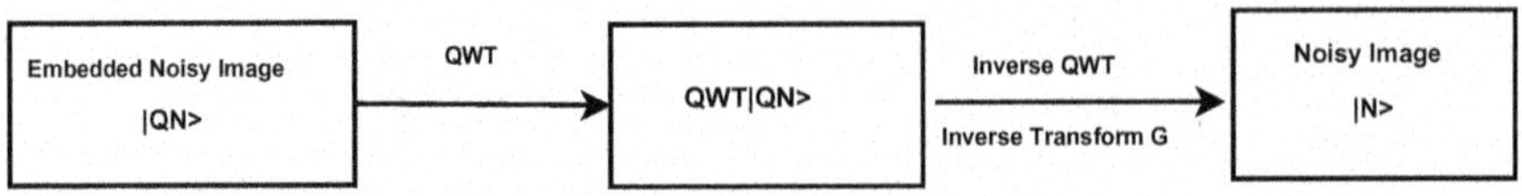

Fig. 9 Block diagram of the overall quantum noisy image extraction procedure

3.6 Noisy Image Extraction

We bring out the noisy image $|N\rangle$ from the embedded image $QWT|QN\rangle$ with the help of inverse QWT(QWT^{-1}) and the inverse transform of the similar geometric position operations G^{-1}. The entire embedding process is reversible because all the internal transforms are unitary in nature. The overall extraction procedure is shown in Fig. 9.

$$\begin{aligned} InverseQWT[QWT(|QN_{out}\rangle)] = QWT^{-1}[QWT(|Q\rangle) \\ +|G_kN\rangle] = |Q\rangle + QWT^{-1}(|G_kN\rangle). \end{aligned} \tag{18}$$

However, the final noisy image can be extracted as

$$|N\rangle = G_k^{-1}[QWT(|QN\rangle) + QWT(|Q\rangle)], \tag{19}$$

where G_k^{-1} is the inverse transform of the G_k.
However, image denoising is an initial process before starting any important operations. However, the measurement process shows output in the form of a probability distribution of the quantum states.

4 Complexity Analysis

In the complexity analysis section, we assume the dimension of the vector is $N = 2^n$. It is already shown in Table 1 that quantum transforms achieve less computational complexity than their classical counterparts. In this process, $2O(2^{4n})$ quantum gates are required for preparing the quantum original image and noisy image, quantum wavelet operation requires $O(n^2)$, and finally, the embedding noisy image process requires overall $O(log^2n)$ operations (including two-point swap geometric transformations) [15]. However, the highest time complexity is $O(2^{4n+1}) = O(2^n)$ but if we neglect the FRQI image preparation process, then the time complexity of the noisy embedding process is $O(log^2n)$ [15]. However, unlike the classical case, the computational time complexity of the quantum thresholding operation using Quantum Oracle is $O(1)$ as it includes the inherent parallelism of quantum computation. So, the total computational time complexity of our proposed technique is $O(log^2n)$. A detailed comparison is shown in Table 2. Table 3 describes the performance com-

Table 2 Comparison with some other popular image denoising methods

Non-local ID [16]	PCA-ID [17]	Curvelet-ID [18]	Adaptive Wavelet-ID [19]	Orthogonal Wavelet-ID [22]	Proposed approach
$O(n^4)$	$O(n^3)$	$O(n^2 log(n))$	$O(n^2)$	O(nlog(n))	$O(log^2 n)$

Table 3 The performance comparisons in terms of storage for a $2^n \times 2^n$ size of image

Algorithms	Storage (grayscale images)	Quantum image models	Complexity of quantum image construction
Proposed approach	$(2n + 8)$ qubits	FRQI	$O(2^{4n})$
Quantum median filtering [11]	$(2n + 8)$ qubits	NEQR	$O(n2^{2n})$
Classical image denoising using wavelet and curvelet [12, 18]	$(2^n \times 2^n \times 8)$ bits	–	–
Classical image denoising using orthonormal wavelet [22]	$(2^n \times 2^n \times 8)$ bits	–	–
Classical cloud noise filtering [21]	$(2^n \times 2 \times 8)$ bits	–	–

parisons between the proposed technique and other denoising algorithms (classical wavelet [12] and cloud noise filtering [21] and quantum median filtering [11]) for an image with a size of $2^n \times 2^n$. In terms of storage complexity, our proposed algorithm requires $2n + 8$ qubits, which has a noticeable decrease than classical algorithms.

5 Conclusions

Image denoising is applied as a prior part of computer vision, pattern recognition, image security, and so on. In this work, we first show how can we form a scrambled noisy image through quantum techniques and then we have used fourth-order Daubechies quantum wavelet kernel and some elementary quantum operations to perform quantum image denoising operation. This method uses a basic quantum thresholding technique to classify the qubits with high coefficients from the low-coefficient qubits. In this article, we have also shown a comparison of our quantum image denoising approach with some classical denoising approaches. It also proved that how our quantum approach outperforms some popular classical approaches in terms of computational time complexity and storage complexity.

References

1. A. Fijany, C.P. Williams, Quantum wavelet transforms: Fast algorithms and complete circuits, in *NASA International Conference on Quantum Computing and Quantum Communications* (Springer, Berlin, Heidelberg, 1998), pp. 10–33
2. X.H. Song, S. Wang, S. Liu, A.A.A. El-Latif, X.M. Niu, A dynamic watermarking scheme for quantum images using quantum wavelet transform. Quantum Inf. Process. **12**(12), 3689–3706 (2013)
3. S. Caraiman, V.I. Manta, Image segmentation on a quantum computer. Quantum Inf. Process. **14**(5), 1693–1715 (2015)
4. P.Q. Le, F. Dong, K. Hirota, A flexible representation of quantum images for polynomial preparation, image compression, and processing operations. Quantum Inf. Process. **10**(1), 63–84 (2011)
5. V. Vedral, A. Barenco, A. Ekert, Quantum networks for elementary arithmetic operations. Phys. Rev. A **54**(1), 147–153 (1996)
6. L. Fan, L. Fan, C.L. Tan, A Diffusion Process for Wavelet-Transform-based Image Denoising
7. G. Gilboa, N. Sochen, Y.Y. Zeevi, Forward-and-backward diffusion processes for adaptive image enhancement and denoising. IEEE Trans. Image Process. **11**(7), 689–703 (2002)
8. S. Yuan, X. Mao, J. Zhou, X. Wang, Quantum image filtering in the spatial domain. Int. J. Theor. Phys. **56**(8), 2495–2511 (2017)
9. D. McMahon, *Quantum Computing Explained* (Wiley, Hoboken, 2007)
10. V.I. Manta, S. Caraiman, Quantum image filtering in the frequency domain. Adv. Electr. Comput. Eng. **13**(3), 77–84 (2013)
11. P. Li, X. Liu, H. Xiao, Quantum image median filtering in the spatial domain. Quantum Inf. Process. **17**(3), 49 (2018)
12. S.G. Chang, B. Yu, M. Vetterli, Adaptive wavelet thresholding for image denoising and compression. IEEE Trans. Image Process. **9**(9), 1532–1546 (2000)
13. S. Chakraborty, S.B. Mandal, S.H. Shaikh, Design and implementation of a multivalued quantum circuit for threshold based color image segmentation. Intell. Decis. Technol. **12**(2), 251–264 (2018)
14. S. Chakraborty, S.B. Mandal, S.H. Shaikh, Quantum image processing: challenges and future research issues. Int. J. Inf. Technol. **10**(3), 1–15
15. P.Q. Le, A.M. Iliyasu, F. Dong, K. Hirota, Fast geometric transformations on quantum images. Int. J. Appl. Math. **40**, 3 (2010)
16. J. Wang, Y. Guo, Y. Ying, Y. Liu, Q. Peng, Fast non-local algorithm for image denoising. In *2006 International Conference on Image Processing*, IEEE, pp. 1429–1432 (2006)
17. G. Chen, S.E. Qian, Denoising of hyperspectral imagery using principal component analysis and wavelet shrinkage. IEEE Trans. Geosci. Remote Sens. **49**(3), 973–980 (2011)
18. J.L. Starck, E.J. Candes, D.L. Donoho, The curvelet transform for image denoising. IEEE Trans. Image Process. **11**(6), 670–684 (2002)
19. Z. Cai, T.H. Cheng, C. Lu, K.R. Subramanian, Efficient wavelet based image denoising algorithm. IEEE Electron. Lett. **37**(11), 670–684 (2001)
20. D.S. Oliveira, R.V. Ramos, Quantum bit string comparator: Circuits and applications. Quantum Comput. Comput. **7**(1), 17–26 (2007)
21. S. Surendhar, P. Thirumurugan, S. Sasikumar, A denoising architecture for removing impulse noise in image. Int. J. Innov. Res. Sci. Eng. Technol. **3**(1), (2014)
22. A.S. Kori, A.S. Manjunatha, An efficient method for image denoising using orthogonal wavelet transform. Int. J. Sci. Res. **4**(3), 2040–2043 (2015)

Semantic Image Completion and Enhancement Using GANs

Priyansh Saxena, Raahat Gupta, Akshat Maheshwari, and Saumil Maheshwari

Abstract Semantic inpainting or image completion alludes to the task of inferring arbitrary large missing regions in images based on image semantics. Since the prediction of image pixels requires an indication of high-level context, this makes it significantly tougher than image completion, which is often more concerned with correcting data corruption and removing entire objects from the input image. On the other hand, image enhancement attempts to eliminate unwanted noise and blur from the image along with sustaining most of the image details. Efficient image completion and enhancement model should be able to recover the corrupted and masked regions in images and then refine the image further to increase the quality of the output image. Generative Adversarial Networks (GAN) have turned out to be helpful in picture completion tasks. In this chapter, we will discuss the underlying GAN architecture and how they can be used for image completion tasks.

1 Introduction to GAN

Generative Adversarial Network (GAN) is a type of deep neural network that is becoming increasingly popular in these days. They have immense applications in doing tasks which were once considered to be too complicated for computers to solve.

P. Saxena (✉) · R. Gupta · A. Maheshwari · S. Maheshwari
ABV-Indian Institute of Information Technology and Management, Gwalior, India
e-mail: saxenapriyanshasd@gmail.com

R. Gupta
e-mail: raahat.gupta.1998@gmail.com

A. Maheshwari
e-mail: aks3d76@gmail.com

S. Maheshwari
e-mail: saumilmaheshwari@yahoo.co.in

A. Nanda and N. Chaurasia (eds.), *High Performance Vision Intelligence*,
Studies in Computational Intelligence 913,
https://doi.org/10.1007/978-981-15-6844-2_11

This chapter first gives an introduction of GAN along with intuitive examples, and then lays out a brief description of its applications followed by one such application in detail—*how to use GAN for the task of image completion.*

After reading this chapter, readers will be able to appreciate the beauty of generative networks and apply then to real-scale applications. Let us first start with a brief overview of the short albeit fascinating history of GAN.

1.1 History

There are several tasks where computers have matched, or even surpassed, human performance. Computers have come a long way and even exceeded human intelligence in fields like face recognition, classification, or even games such as *Go*. However, there are still many applications where computers have a long way to go—the most prominent among them being chatbots or voice-powered assistants like Alexa, Google Home, etc. One can easily distinguish a conversation with a voice assistant versus a human. Thus, we can say that computers have not entirely passed the Turing test yet.

One of the main reasons for this "gap" of intelligence is that historically A.I. algorithms were remarkable at *learning* various complex patterns and structures among data, and using them to make a decision, such as a classification or decision the next move in *Go*. However, computers were not so good in *generating* new data—something that humans have to learn from a very early stage.

This "gap" was significantly reduced when Ian Goodfellow et al. published the paper on **Generative Adversarial Networks (GAN)** in 2014 [1]. While we cannot say that GANs were the first algorithms to enable computers to *generate* new data—other algorithms have existed before them—but GANs were able to do the task significantly better than others and at degrees that matched production levels.

So, we can say that GANs enable us to accomplish milestones of artificial intelligence that were considered near-impossible, even just before the time the paper was published. Abilities of GAN include generating fake images in a real-world environment, removing objects from images, converting a video clip of a horse into a zebra, and many more! GANs are so remarkable that even industry experts like Yann LeCun, Director of AI research at Facebook, were quoted as saying that GANs and their variations are "the coolest idea in deep learning in the last twenty years."[1]

1.2 But, What Exactly Is a GAN?

In a nutshell, **Generative Adversarial Networks** use an aggressive, game-like environment to train two neural networks. The networks "compete" with each other, and

[1]LeCun, Yann. "RL Seminar: The Next Frontier in AI: Unsupervised Learning".

the end result is an image or a sequence of words that is indistinguishable from what would appear in the real world.

GAN comprises two networks—the **generator** and the **discriminator**. The job of the generator is to manufacture data (images or text) that is indistinguishable from the real examples in the training set. The discriminator, on the other hand, is tasked with identifying the data coming from the generator versus (the *fake* examples) the data in training set (the *real* data).

The generator, on its way to produce realistic-looking images, receives feedback from the discriminator. The goal of the generator is to fool the discriminator as many times as possible. This gives it a metric to measure its performance, and this metric can effectively be used to train the generator.

The discriminator trains as well—it is given feedback based on the percentage of images correctly classified as fake versus the fake images that get away. Its goal is to classify as much fake images as possible, and so in a sense we can say that *the generator and discriminator compete with each other* when a GAN is training. Both networks continue to improve as this cat-and-mouse game progresses.

One of the curious properties of GAN—and one which makes them particularly hard to train—is that the optimization minimum is not fixed in GAN. Normally, gradient descent contains an optimization function which tires to minimize a certain *cost function*. However, in case of GAN, the optimization function is seeking a balance between the two opposing forces; it continues to work until a state of *equilibrium* is achieved.

1.3 An Intuitive Example

Let us picture an analogy to understand the inner workings of GAN better. Consider a small town which has an organized crime unit—the local mafia. The mafia tries to counterfeit money, and every time it produces a new batch of bills, it attempts to deposit them in the bank (via an associate, of course).

If the associate gets arrested, the mafia learns that the bank has judged their bills as counterfeit and knows that it has to improve in the production of bills. Just like a *generator* whose examples are rejected, the mafia does not know exactly where it went wrong, just that its attempt to fool the bank was not fruitful. It goes back to work, and its team of chemists and analysts produce another set of bills, probably better than the previous ones.

Similarly, the bank gets better at distinguishing counterfeit bills over time. It acts as the *discriminator* and provides *feedback* to the generator at the end of each iterator (by arresting their associate or not). The bank may decide to invest in better money scanning technologies, hire experts, etc. so that no counterfeit money gets by.

In this manner, the generator and discriminator both get to improve, and we are left with a bunch of high-quality counterfeit money at the end!

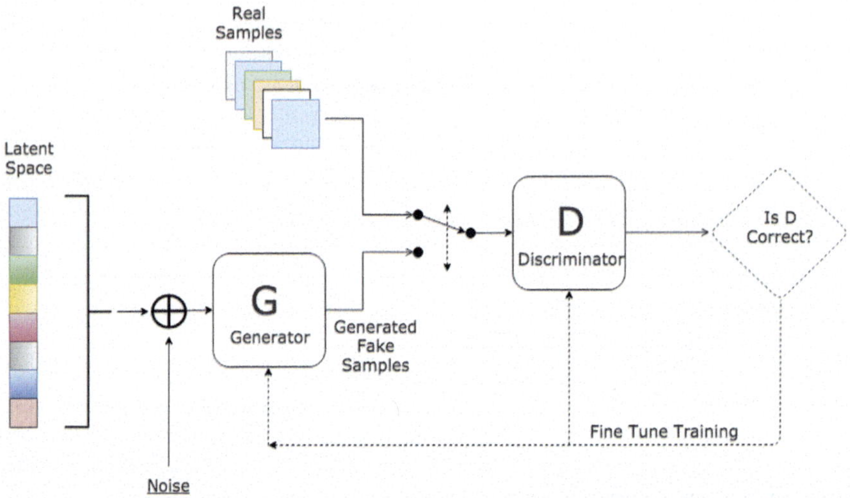

Fig. 1 The basic architecture of generative adversarial networks (*Source* https://www.linkedin.com/pulse/gans-one-hottest-topics-machine-learning-al-gharakhanlau/?trk=pulse_spock-articles)

2 GANs in Action

Now that everything is probably clear about the basic workings of GAN, we look at the specific architecture and terms associated with GANs in this section. The pseudocode of GAN training algorithm along with visualization is also described. We end this section by giving various applications where GAN is used in real life.

2.1 Architecture of GAN

Imagine our goal is to teach a GAN to produce realistic-looking handwritten digits. Figure 1 shows the diagram of core GAN architecture.

The architecture can be classified into the following parts:

1. **Training Dataset**: This is the set of real images shown in the upper left corner of the image. The real and labeled examples are shown only to the discriminator and not to the generator. Thus, the latter has to generate counterfeit images which seeing what an actual one looks like. The inputs are generally represented by x.
2. **Random Noise Vector**: The input to the generator also contains a vector of random numbers (z) in addition to the latent space as shown in the left side of the image. The generator uses the random noise as a starting point of its synthesis process.

3. **Generator Network**: It is represented by G, it takes the random numbers z and creates a set of fake examples $x*$. Its goal is to make these fake examples as indistinguishable from the real ones as possible.
4. **Discriminator Network**: The discriminator D inputs both x and $x*$, and gives a score (on a scale from 0 to 1) based on the identicality of these inputs.
5. **GAN Training**: For each of the discriminator's output, its performance is determined with respect to a predefined metric. This performance is used to update both generator and discriminator through backpropagation:

   a. Discriminator's weights and biases get updated to maximize its classification accuracy;
   b. Generator's weights and biases get updated to maximize the likelihood that the discriminator incorrectly classifies $x*$ as real.

2.2 GAN Training Algorithm

As GAN consists of two entities, discriminator and generator, the depiction of steps for both during training of GAN [2] is given below:

The Discriminator

1. We take a random sample from the training dataset. Label it as x.
2. Use the generator network to produce a batch of fake images $x*$ (note this step includes taking a random noise vector z first).
3. The discriminator is fed inputs x and $x*$; it is used to classify among the two.
4. The classification errors and losses are computed, and, through the process of backpropagation, the weights and biases of discriminator are updated accordingly. *The goal of discriminator is to minimize the classification errors.*

The Generator

1. We take a new random noise vector z.
2. The generator uses z as input to produce a batch of fake images $x*$.
3. This $x*$ is fed to the discriminator, which calculates the classification score based on x and $x*$.
4. The error and loss are computed and, via backpropagation, the weights and biases of generator are updated accordingly.
   The goal of generator is to maximize the classification errors.

   The pseudocode of the GAN training is depicted in Algorithm 1.

Algorithm 1: GAN Training Algorithm

for each training iteration
  Train the Discriminator
    Take a random sample from the training set. Label it as x.
    Take a new random noise vector. Label it as z.
    Utilizing the Generator, use z to manufacture a counterfeit example $x*$.
    Use the Discriminator and classify x and $x*$.
    Compute classification losses and use backpropagation for updating the discriminator weights and biases. Minimize classification errors.
  Train the Generator.
    Take a new random noise vector z.
    Utilizing the Generator, use z to manufacture a counterfeit example $x*$.
    Use the Discriminator and label $x*$.
    Compute classification losses and use backpropagation for updating the generator weights and biases. Maximize Discriminator errors.
end **for**

2.3 *When to Stop Training in GAN?*

As stated earlier, the optimization minimum is not fixed in the case of GAN. Our goal is to seek a balance between the two opposing forces—the generator and the discriminator. This could create a state of ambiguity as to when to stop training. *How can we be certain that the state of equilibrium is achieved, and further training will not necessarily benefit the GAN algorithm?*

This problem is somewhat similar to the *zero-sum game* in Game Theory—in this case, one player's gains come from another's losses, and the exact value that one wins, the other loses. All zero-sum games include a point where neither player can improve their situation by changing any of their actions and such a point is called *Nash Equilibrium*.[2] Our goal is to find the Nash equilibrium in case of GAN models.

The generator and the discriminator reach their Nash equilibrium when the fake examples produced by the former are indistinguishable from the real data, and the latter can, at best, randomly guess whether a particular example is real or fake. Let us look at the following two cases:

- Although it may seem that the discriminator is just guessing real and fake at random 50% probability (and hence has room to improve), the fake examples $x*$ are truly indistinguishable from the real examples x, and hence there is nothing the discriminator can do to tell them apart from one another. Thus, it cannot do better than a random chance probability.
- The generator is likewise at a point where it has nothing to gain from further tuning. Because the examples it produces are already perfectly indistinguishable from the

[2]Nash Equilibrium is named after the American economist and mathematician John Forbes Nash Jr, whose life and career were captured in the biography titled *A Beautiful Mind* and inspired the eponymous film.

real ones, even a tiny change to the process it uses to turn the random noise vector z into a fake example $x*$ may give the discriminator a clue about how to tell apart the fake example from the real data, making the generator worse off.

A GAN is fully trained when Nash equilibrium is achieved. But, as if often the case with such optimization problems, such an equilibrium is *very* hard to reach in practice. We stop as soon as we achieve our desired objective at some judgement level.

2.4 Applications of GAN

Although it is almost impossible to list the entire applications of GAN in this section, a few important and interesting ones are listed here:

- Generating photo-realistic fake images: Take a look at Fig. 2. All the faces look real, aren't they? But the fact is none of the faces are real; they're all produced using *Progressive Growing* of GAN [3].
- Image-to-Image Translation: We can have a little fun and use an implementation called *CycleGAN* to replace a horse in an image with a zebra (and vice versa) while keeping all other factors same (as shown in Fig. 3). We can also do something more meaningful and convert a photograph into a monet [4].
- Automatic synthesis of realistic images from a textual sentence using *Stack GAN*, and transferring style from one domain to another domain using *Discovery GAN* [5].
- Generating realistic image from attributes: Imagine that burglar breaks in your house at night. You catch a brief glimpse of him/her, but nothing useful enough to identify the person. Now, suppose there's an advance system in the police station that could generate a realistic image of the thief based on the description provided by you. That system would probably use GAN [6].

3 Image Completion

This section will discuss about one of the most common and highly interesting applications where Generative Adversarial Networks are used: the task of **Image Completion**.

Image completion, or more formally, *semantic inpainting* defines the task of completing arbitrary-sized unknown subsets of images based on semantics [7]. Our task is primarily to predict accurately high-level useful content; this is considerably difficult than just filling in the images with spurious data. Semantic implantation finds applications in restoration of damaged artworks, editing images to remove inanimate objects from them, etc. An interesting range of applications and results is shown in sections below.

Fig. 2 Generating fake faces using GAN [3]

Fig. 3 Converting horses to zebras and monets to photos using GAN [4]

3.1 Related Works

Jia-Bin Huang and Ahuja have proposed an advance-knowledge approach which used contextual information for image completion. However, in case the corrupted region is large or is irrelevant to visual data, or if the complexity of the image is high, the output of the method would be quite unsatisfactory [8]. Connelly Barnes and Eli have proposed patch matching algorithm for image completion for nonparametric texture construction. The algorithm performed satisfactorily and was able to identify similar patches. However, it failed when the original image lacked adequate data to complete the missing regions [9]. Yunjin Chen and Thomas Pock have proposed nonlinear response dispersion model, which consists of a feedforward network with a fixed number of gradient descent stages. Trainable nonlinear reaction–diffusion accomplished promising execution in any case; its display was prepared for a specific noise level. It was unfit to perform well on pictures with obscure noise levels. Additionally, it requires the output which is expected by the network during training [10]. In 2011, Deng transformed the inpainting task to the graph-labeling task using graph Laplace method. However, this method required image samples of the image to be inpainted be included in the training data, which was not practical in real-life applications [11]. A viable face inpainting algorithm utilizing a generative model was proposed by Yijun Li, Sifei Liu, and Jimei Yang. From background inpainting task, face inpainting is a challenging task because it regularly needs to produce semantically newer pixel areas in the missing region parts like eyes and nose, which can vary from person to person. Even though the model had the capacity to produce semantically conceivable and outwardly satisfying content, it has a few constraints. The model still could not deal with some unaligned faces. Also, it did not wholly misuse the spatial connections between nearby pixels [12]. Ruijun and Yang proposed an improved generative translation model. The paper proposed a semantic image completion method using regional completions for painting completion. Using the generator and discriminator network, the missing region is generated, which should be consistent with the surrounding region. However, image completion work is restricted to only face data and needed to be improved to ensure that the entire painting work could be recovered [13].

Deepak Pathak puts forward Context Encoders (CE) which estimated missing areas in images based on its surroundings. However, during training it needed a mask on the corrupted regions of the image, that is, a significant disadvantage of the approach, and also context encoders led to blurry and noisy results in the inpainted parts [14]. Ren proposed a novel CNN architecture named Shepard convolutional neural networks which efficiently equip conventional CNN with the ability to learn missing data. However, in case the corrupted region was large or was irrelevant to visual data, or if the complexity of the image is high, the output of the method would be quite unsatisfactory [15].

In [16], low-light enhancement model using convolutional neural network and Retinex theory were proposed. It showed an equivalence between multi-scale Retinex and feedforward convolutional neural network using Gaussian kernels. However, because of the limited receptive field in their model, very smooth regions such as clear sky are sometimes attacked by the halo effect. Jeremias sulam formulated trainlets to construct large adaptable atoms using various datasets of facial images using dictionary learning algorithm. Because of the computational constraints, this method was applied to tiny regions of the image and not on the entire image. As a result, this approach did not give satisfactory results on large regions in images [17]. Raymond and Chen [18] proposed another picture completion technique that can be utilized to fix any state of gaps. In any case, such training depends on the data used in training. In the meantime, the processing of surface and structure was not sufficiently impeccable. Kai Zhang and Yunjin Chen proposed a picture denoising approach in which they built feedforward denoising convolutional neural systems using residual learning and batch normalization. However, this methodology was unfit to recover missing regions, and it just denoised the picture. Likewise, it was unfit to refine pictures with genuine complex commotion and other general picture restoration tasks [19].

4 Introducing Wasserstein GAN for Image Completion

Now that you've known what image completion really is, let us look at how it can be solved with GANs. Though other GAN architectures are available, we are using the one called **Wasserstein GAN** for this purpose. Wasserstein GAN is an architecture that can be used for image completion tasks. It creates the coarse patches to fill the missing region in the distorted picture, and the enhancement network will additionally refine the resultant pictures utilizing residual learning procedures and hence give better complete pictures for computer vision applications. The algorithms are described in the following sections. For an overview of the results in the CelebA-hq dataset, you can go directly to Sect. 6.

4.1 Methodology

The methodology could be separated into three different steps.

In the first step, data preprocessing on CelebA-hq dataset[3] is done to train and test the developed model.

The following data preprocessing steps were followed:

- The dataset is splitted into 15,000 training images and around 1000 testing images.
- Each face image in the dataset is resized to 64* 64* 3 pixels to train the Wasserstein GAN model.
- Masking—a binary mask is used with values 0 or 1, where 0 corresponds to the corrupted region and 1 corresponds to the uncorrupted region in the image. This binary mask is applied to all images to make them corrupted which will serve as input of the training process.

In the second step, a Wasserstein-GAN-based model to complete the missing pixels in the image is developed. The image completion GAN gives a complete image with a blurry filled area. The generator of the GAN generates real looking images, but in the process of generation, the noise gets unavoidably added.

So, in the third step, the output of the generator is passed through the enhancement network to make the filled area clear and to refine the completed image further. The enhancement network is trained using 2000 image pairs containing blurry images and its corresponding clean images.

---

[3] http://mmlab.ie.cuhk.edu.hk/projects/CelebA.html.

4.2 Wasserstein Distance as Loss

This architecture is different from the one proposed by Goodfellow [1] in that it uses Wasserstein distance as to train the generator so that it can capture training data distribution and generate images similar to those in the training data.

Wasserstein distance is a measure of the distance between two probability distributions. For the generated data distribution p_g and the real data distribution p_r, it can be mathematically defined as the cost for the cheapest plan from p_g to p_r. It is also called critic loss or Wasserstein distance. The Wasserstein distance loss function L to train the generator can be mathematically represented as follows:

$$L = \mathbb{E}_{\tilde{x} \sim p_g} |C(\tilde{x})| - \mathbb{E}_{x - p_r} |C(x)| \quad (1)$$

Here, the first term represents the expectation of the distribution generated by the generator and the second term represents the expectation of the real training data distribution. By minimizing the difference between the two, the generator learns to generate samples having probability distribution similar to training data distribution. Now, to make the learning faster and make model convergence faster, gradient penalty term is added to our loss function. So, the overall loss function L of Wasserstein GAN becomes

$$L = \mathbb{E}_{\tilde{x} \sim p_g} |C(\tilde{x})| - \mathbb{E}_{x - p_r} |C(x)| + gradient\ penalty \quad (2)$$

where gradient penalty will be given by

$$gradient\, penalty = \lambda \underset{\hat{x} \sim P_g}{\mathbb{E}} [(|| \nabla_{\hat{x}}\ C(\hat{x})||_2 - 1)^2] \quad (3)$$

Here λ is the gradient penalty coefficient.

4.3 Image Generation Using Wasserstein GAN

After training the generator to generate samples which look real, the next aim is to ensure that the missing region generated has a similar context to the non-missing region so that sensible looking completed images as output can be obtained.

A binary mask with values 0 or 1 is used. 0 corresponds to the corrupted region while 1 corresponds to the uncorrupted region in the image. Let y represent the uncorrupted image. $M \odot y$ gives the uncorrupted part of the image. Let $G(z')$ be some image generated by the generator which suitably completes the missing region in the image. $(1 - M) \odot G(z')$ represents the completed region which when added to the uncorrupted region gives the reconstructed image as output [20]:

$$x_{reconstructed} = M \odot y + (1 - M) \odot G(z') \quad (4)$$

To find z' that suitably completes the image following loss functions are defined: **Contextual Loss:** To ensure both generated and the input images have same context, ensure that the uncorrupted pixel in original image y is same as the pixels in the generated image G(z) at a particular location. For this, pixel-wise difference between the uncorrupted part of the two images is taken and then this difference is minimized.

$$L_{contextual}(z) = ||M \odot G(z) - M \odot y||_1 \tag{5}$$

where $||x||_1$ represents l_1 norm of some vector x.

Perceptual Loss: It ensures that the output image looks real. For this, the perceptual loss is given below:

$$L_{perceptual}(z) = log(1 - C(G(z))) \tag{6}$$

Total Loss: It is a sum of perceptual and contextual loss and is denoted by L(z):

$$L(z) = L_{contextual}(z) + QL_{perceptual}(z) \tag{7}$$

where Q is a hyperparameter and we minimize this loss function to ensure the completed image is contextually similar to the input image.

4.4 Enhancement Network

In enhancement network to refine completed images, the residual learning approach is used. The input to the network is blurry image y = x + v, where x is the clear image and v represents the blur added. The residual network is trained to grasp the mapping R(y) ≈ v, to get the clear image x as x = y − R(y). Mathematically, the average mean square error among the output residual image by the model and the actual residual images is used as error function for getting the parameter Θ to train the enhancement network.

$$L(\theta) = \frac{1}{2N}\sum_{i=1}^{N} ||R(y_i;\theta) - (y_i - x_i)||^2 \tag{8}$$

Here, L is the training error of the enhancement network and N are total training images. Enhancement network consists of following layers as shown in Fig. 4: (i) Conv+ReLU: It creates feature maps, and ReLU adds the non-linearity. (ii) Conv+BN+ReLU: This layers contains added batch normalization between Conv and ReLU. (iii) Conv: It is used to get the output residual image.

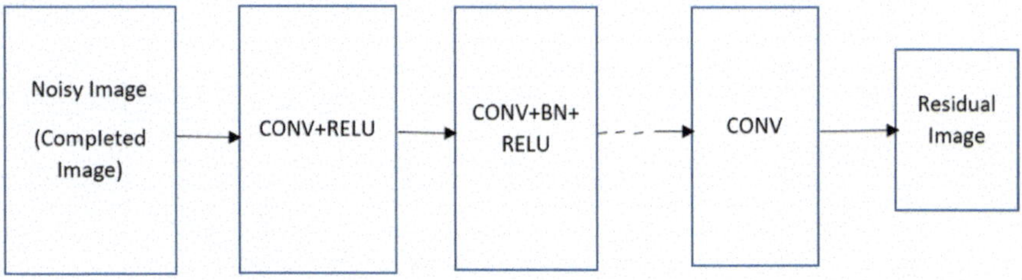

Fig. 4 Enhancement network to refine completed images

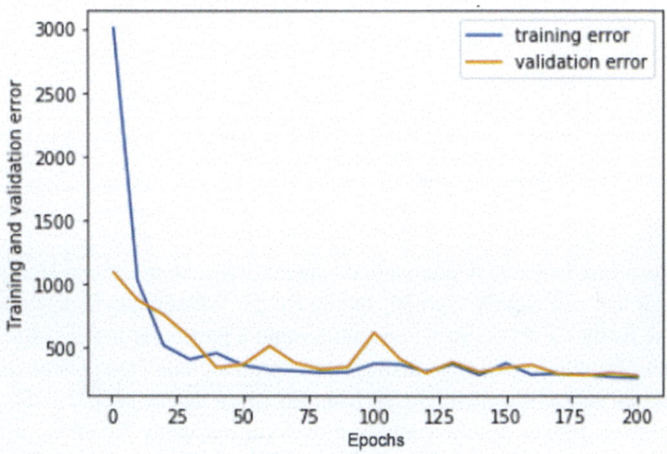

Fig. 5 Enhancement network training plot to refine completed images

4.5 Results and Discussion

The following plot was obtained by training the enhancement network on 2000 celeba-hq image pairs of clean and its corresponding blurry images.

In Fig. 5, we observe that, as the training proceeds, the average mean square error among the output residual image by the model and the actual residual images decreases. As a result, according to Eq. (8), the training error decreases. Finally, around 200th epoch, the enhancement network is sufficiently trained ,which is evident as the training error becomes constant at a particular value, and there is no further decrease.

The Wasserstein GAN model is trained on 15,000 Celeba-hq images for 10,000 epochs and batch size of 128. It is seen that in the initial stages of learning the expectation of the distribution generated by the generator is different from the expectation of the distribution of real data and hence the difference between the two is higher resulting in higher Wasserstein distance values. However, as the learning proceeds generator learns the distribution of the real data and then generates samples having a similar distribution with the real data, and hence the difference in their expectation

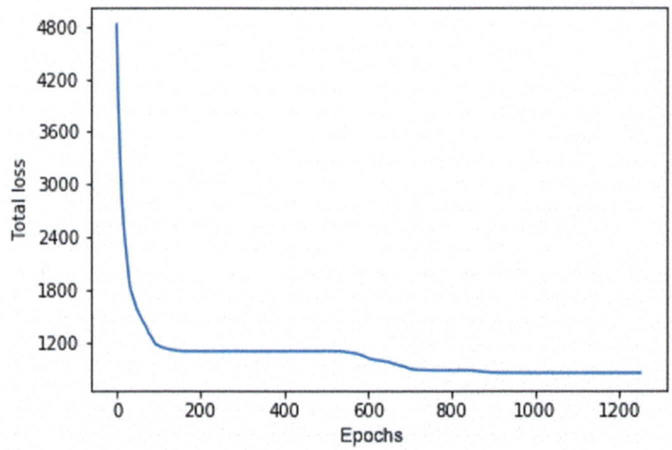

Fig. 6 Total image completion loss plot

decreases resulting in lower Wasserstein distance values. Now, around 10,000 epochs the generator has sufficiently learned, and hence the Wasserstein distance values do not decrease further and becomes constant around a particular lower value.

In the initial stages, the context in the uncorrupted region of the generated samples and the original samples is different, so from Eq. (5), it can be seen that the resulting contextual loss is higher. As the training moves further using Adam's optimizer (z) gets trained, and hence there is a significant decrease in the contextual loss values. Around 1200th epoch, the context in the uncorrupted region of the generated samples and the original samples becomes quite familiar, and hence the contextual loss becomes constant around a particular value.

Initially, the distribution of generated images and real images is different, so the critic is able to distinguish the generated samples from the real ones and hence the value of C(G(z)) is close to 0 and as a result 1-C(G(z)) becomes close to 1; as a result from Eq. (6), the loss is higher. However, as learning proceeds, the generator generates real looking samples as a result C(G(z)) becomes close to 1 and 1-C(G(z)) becomes close to 0, resulting in lower perceptual loss values from Eq. (6).

The perceptual loss values are quite lower compared to contextual loss values, and as a result from Eq. (7), the total image completion loss is almost equal to contextual loss, and as a result Fig. 7, which is total image completion loss plot, is almost similar to contextual loss plot for image completion (Fig. 6).

The following two evaluation metrics were used to evaluate the quality of output images by the model.

4.6 Peak Signal-to-Noise Ratio (PSNR)

PSNR [21] is measured in decibels (dB). The higher the PSNR, the better image has been completed to match the original image.

$$MSE = \frac{1}{mn}\sum_{i=0}^{m-1}\sum_{j=0}^{n-1}||f(i,j) - g(i,j)||^2 \tag{9}$$

$$PSNR = 20\log_{10}\left(\frac{MAX_f}{\sqrt{MSE}}\right) \tag{10}$$

Here, f is the original image, g represents completed image through the model, m represents image pixel rows, n represents image pixel columns, and i and j represent row and column indexes, respectively. MAX_f is a constant equal to 255.

5 Structural Similarity Index (SSIM)

The Structural Similarity (SSIM) Index [21] depends on computation of terms, namely, the luminance, contrast, and structural term.

$$SSIM(x, y) = [C(x, y)]^{\alpha} \times [I(x, y)]^{\beta} \times [S(x, y)]^{\gamma} \tag{11}$$

where C represents contrast, I represents luminance, S represents structural term, x represents original, and y represents completed images. The parameters $\alpha > 0$, $\beta > 0$, and $\gamma > 0$ are used to adjust the relative importance of the three components.

The following *PSNR* and *SSIM* values through the proposed approach are compared with the existing techniques in Table 1.

It can be seen that the approach performs well in CelebA-hq dataset compared to other proposed image completion techniques, which is evident from the above PSNR and SSIM values.

Some of the results obtained through the proposed image completion approach using Wasserstein GAN are shown below in Fig. 7.

Table 1 Comparison of PSNR values

Methods	CE [14]	PI [22]	Proposed approach
PSNR (dB)	22.85	21.45	23.41
SSIM	0.872	0.851	0.9074

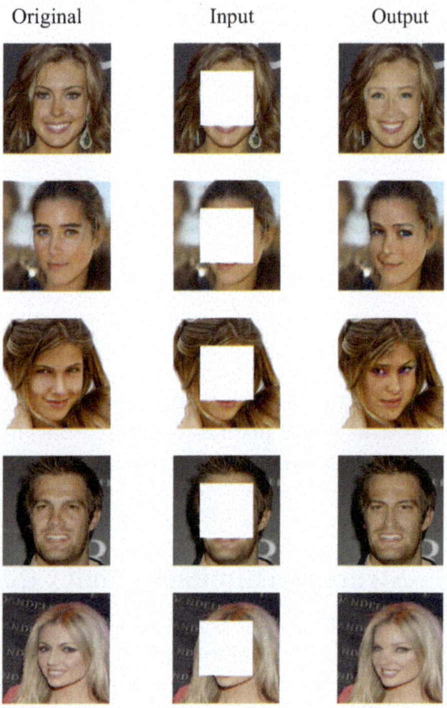

Fig. 7 Experimental results

6 Applications of Image Completion Using GAN

This section looks at some of the real-world applications where GANs are useful. As depicted by the results in Figs. 8, 9, and 10, GAN does an astonishing job and the generator images are mostly alike to the actual/probable images.

Portrait Completion Sometimes parts of a portrayal are missing from the image, due to the image being cropped, an opaque article blocking our object of interest, etc. GANs are used in this scenario to complete the portrait as if the blocking article was never there [23]. Figure 8 depicts the results obtained.

Sunglasses Removal Not only are sunglasses worn to look more attractive, they can also hide the identity of lawbreakers and suspects. Thus, sunglasses removal is an active problem used by law enforcement to identify sunglasses-wearing suspects in surveillance footage with their known photographs. Results of the same are depicted in Fig. 9.

Object Removal on Face Similar to the above problem, it also involves removing an inanimate or unwanted object on the face. Examples include removing that headband that went out of fashion years ago, removing a ring from one's finger, etc. Figure 10 shows impressive results in this domain.

Fig. 8 Portrait completion using GAN [23]

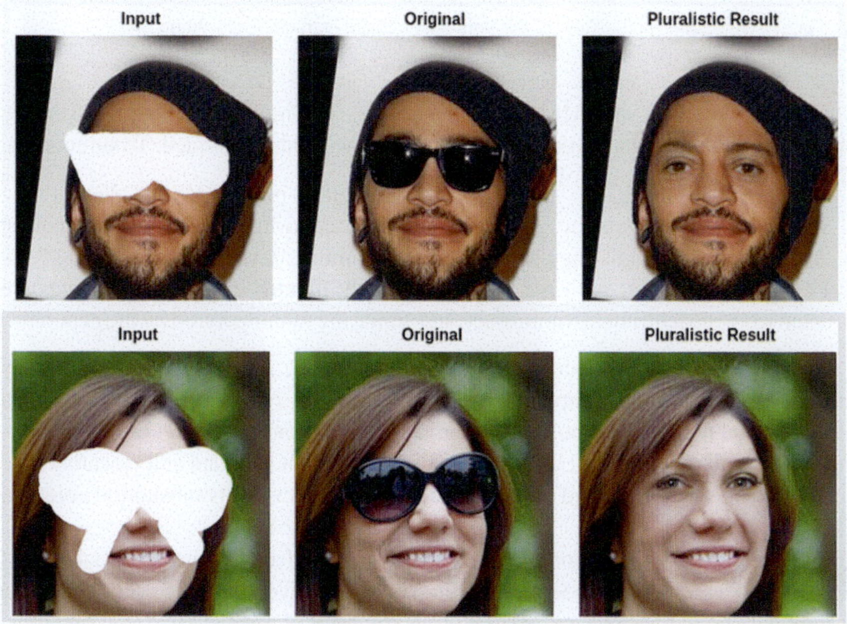

Fig. 9 Sunglasses removal using GAN (*Source* http://www.chuanxiaz.com/project/pluralistic/)

Fig. 10 Headband removal using GAN (*Source* http://www.chuanxiaz.com/project/pluralistic/)

7 Summary

This chapter introduced GANs and their working. The intuition behind the working of GAN and the relevant mathematics of the topic were also provided. To summarize, GANs are made up of the following unit:

- **The Generator:** with the goal of fooling the discriminator by manufacturing instances that are indistinguishable from the training dataset and
- **The Discriminator:** with the goal of correctly identifying which instances are taken from the legitimate training set, and which are manufactured by the generator.

GANs can be used to create hyperrealistic imagery, i.e., imagery that is completely artificially generated by computers but looks completely real to the human eyes. On a philosophical note, all technological innovations have misuses. And since it is seemingly impossible to uninvent a technology, it is the job of scholars and researchers to keep this technology safe and work actively toward achieving its substantial potential [2].

In this chapter, we were only able to scratch the surface of what is possible with GANs; however, we hope that after reading this chapter you have the necessary basic theoretical knowledge to continue exploring any facet of this field that you find most interesting.

8 Further Study

- Deep Learning with Python: Francois Chollet.
- GANs in Action: Jakub Langr and Vladimir Bok.
- Learning General Adversarial Networks: Kuntal Ganguly.
- Online image completion demo by Chuanxia Zheng et al. at http://www.chuanxiaz.com/project/pluralistic.

References

1. I. Goodfellow et al., Generative adversarial nets, in *Advances in Neural Information Processing Systems* (2014)
2. J. Langr, V. Bok, *GAN in Action*
3. T. Karras et al., Progressive growing of GANs for improved quality, stability, and variation (2017), arXiv:1710.10196
4. J.-Y. Zhu et al., Unpaired image-to-image translation using cycle-consistent adversarial networks, in *Proceedings of the IEEE International Conference on Computer Vision* (2017)
5. H. Zhang et al., StackGAN: text to photo-realistic image synthesis with stacked generative adversarial networks, in *Proceedings of the IEEE International Conference on Computer Vision* (2017)
6. S. Reed et al., Generative adversarial text to image synthesis (2016), arXiv:1605.05396
7. R.A. Yeh et al., Semantic image inpainting with deep generative models, in *Proceedings of the IEEE Conference on Computer Vision and Pattern Recognition* (2017)
8. J.-B. Huang, N. Ahuja, Image completion using planar structure guidance. ACM Trans. Graph. (Proc. SIGGRAPH) **33**(4) (2014)
9. C. Barnes, E. Shechtman, A. Finkelstein, D.B. Goldman, PatchMatch: a randomized correspondence algorithm for structural image editing. ACM Trans. Graph. (Proc. SIGGRAPH) **28**(3) (2009)
10. Y. Chen, T. Pock, Trainable nonlinear reaction diffusion: a flexible framework for fast and effective image restoration. IEEE Trans. Pattern Anal. Mach. Intell. **39**(6), 1256–1272 (2017)
11. Y. Deng, Q. Dai, Z. Zhang, Graph Laplace for occluded face completion and recognition. IEEE Trans. Image Process. **20**(8), 2329–2338 (2011)
12. Y. Li, S. Liu, J. Yang, M.-H. Yang, Generative face completion. CoRR (2017), arXiv:1704.05838
13. R. Liu, R. Yang, S. Li, Y. Shi, X. Jin, Painting completion with generative translation models. Multimed. Tools Appl. (Springer) 1–14 (2018)
14. D. Pathak, P. Krahenbuhl, Context encoders: feature learning by inpainting, in *Proceedings of the IEEE Conference on Computer Vision and Pattern Recognition* (2016), pp. 2536–2544
15. J.S.J. Ren, L. Xu, Q. Yan, W. Sun, Shepard convolutional neural networks. Adv. Neural Inf. Process. Syst. **28**, 901–909 (2015)
16. L. Shen, Z. Yue, F. Feng, Q. Chen, S. Liu, J. Ma, MSR-Net: low-light image enhancement using deep convolutional network. CoRR (2017), arXiv:1711.02488
17. J. Sulam, M. Elad, Large inpainting of face images with trainlets. IEEE Signal Process. Lett. **23**(2), 1839–1843 (2016)
18. R.A. Yeh, C. Chen, T.-Y. Lim, M. Hasegawa-Johnson, M.N. Do, Semantic image inpainting with perceptual and contextual losses. CoRR (2016), arXiv:1607.07539
19. K. Zhang, W. Zuo, Y. Chen, D. Meng, L. Zhang, Beyond a Gaussian denoiser: residual learning of deep CNN for image denoising. IEEE Trans. Image Process. **26**, 3142–3155 (2017)

20. C. Chen, T.-Y. Lim, R.A. Yeh, Semantic image inpainting with deep generative models, in *Proceedings of the IEEE Conference on Computer Vision and Pattern Recognition* (2017)
21. A. Hore, D. Ziou, Image quality metrics: PSNR vs SSIM, in *In 2010 20th International Conference on Pattern Recognition* (2010), pp. 2366–2369
22. Y. Chen, H. Hu, An improved method for semantic image inpainting with GANs: progressive inpainting. Neural Process. Lett. (Springer) 1–13 (2018)
23. X. Wu et al., Deep portrait image completion and extrapolation. IEEE Trans. Image Process. (2019)
24. C. Zheng, T.-J. Cham, J. Cai, Pluralistic image completion, in *Proceedings of the IEEE Conference on Computer Vision and Pattern Recognition* (2019)
25. M. Bertalmio et al., Image inpainting, in *Proceedings of the 27th Annual Conference on Computer Graphics and Interactive Techniques* (ACM Press/Addison-Wesley Publishing Co., 2000)
26. I. Goodfellow, M. Mirza, J. Pouget-Abadie, Generative adversarial nets, in *International Conference on Neural Information Processing Systems* (2014), pp. 2672–2680
27. H. Ren, J. Lee, M. El-khamy, DN-ResNet: efficient deep residual network for image denoising. CoRR (2018), arXiv:1810.06766
28. Z. Liu, P. Luo, X. Wang, X. Tang, Deep learning face attributes in the wild, in *Proceedings of International Conference on Computer Vision (ICCV)* (2015)
29. G. Peyr, Manifold models for signals and images. Comput. Vis. Image Underst. **113**(2), 249–260 (2009)
30. M. Arjovsky, S. Chintala, L. Bottou, Wasserstein GAN, Courant Institute of Mathematical Sciences Facebook AI Research (2017)
31. G. Zhao, J. Liu, J. Jiang, W. Wang, A deep cascade of neural networks for image inpainting, deblurring and denoising. Multimed. Tools Appl. **77**(22), 29589–29604 (2018)

A Fusion of Visible and Infrared Images for Victim Detection

Madhuri Gupta

Abstract A disaster is a serious disruption occurring over a relatively short period of time. It can cause a major risk to life of human who are directly affected and those who implicated in the urban search and rescue process. With the advancement of technology and machine learning techniques, the automatic victim detection systems are replacing human beings. These systems employ essential sensors like visual cameras, infrared (IR) images, etc. to capture the information. We propose information fusion from visible and infrared images for automatic detection of victims. In the research work, IR images are primarily used for extracting local feature descriptors and visible images are used for the purpose of skin detection. Skin detection is performed using a non-parametric histogram-based trained skin pixel likelihood model. Three proven local feature descriptors- HOG (Histogram of Oriented Gradient), SURF (Speeded Up Robust Features), and SIFT (Scale Invariant Feature Transform) have been studied in the context of human detection for IR images. Preliminary results indicate that victim detection was better in infrared images than skin color detection in visible images. In case of fusion approach, it was observed that fusion of HOG + skin color detection performed better than the other two combinations. The results are encouraging for human detection using multimodality sensing.

M. Gupta (✉)
Computer Engineering and Information Technology, ABES Engineering College (Affiliated with Abdul Kalam Technical University, Lucknow), Campus 1, 19th KM Stone, NH 24, Ghaziabad, Uttar Pradesh, India
e-mail: madhuri.gupta@abes.ac.in

A. Nanda and N. Chaurasia (eds.), *High Performance Vision Intelligence*,
Studies in Computational Intelligence 913,
https://doi.org/10.1007/978-981-15-6844-2_12

1 Introduction

Environment of our planet has very dynamic phenomena. Number of natural procedures of atmosphere turned out to be perilous to the living life. These natural procedures are termed as Disaster. Disaster is an occurrence emerging with practically no warning that causes serious interruption throughout life and demise or damage to numerous persons [1]. Some natural disasters like Floods, Droughts, Hurricanes, and Earthquakes have ended up being most dangerous which may have injured and killed a large number of people. As indicated by the CRED [2], in 2013 alone the aggregate number of disaster events adds up to around 334 events that results in 97 million influenced and 22,616 died people which is a major disadvantage for human mankind that need to be rescued.

Each disaster situation is distinctive in its own particular manner and displays new and uncommon difficulties to victims and rescue workforces. A primitive task for the rescue services is to evaluate whether there are human victims in the event, as the upshot of this evaluation determines directly the crisis management teams will be disposed to take themselves into. However, in many cases, one needs to come inside the danger zones just to build this assessment. To this end, some teleoperated mobile robots have been set into service worldwide, allowing a human machinist to search for victims from a distance [3]. However, with the recent advance in autonomous robotics, this work could be automated furthermore, if there would be a stead-fast way to automatically detect human victims, even in complicated outdoor illumination environments. Although, there is a problem during the night time. It is not possible to see victim in visible spectrum during poor illumination and other visual constraints such as dust, smoke, fog, etc. that scatter light [4, 5].

The infrared imaging-based sensors can be utilized to cope up with the above-discussed problem. These sensors identify the thermal signatures. But there are also some limitations with thermal sensors like Occlusions, non-uniform thermal signatures, and low contrast ratio with background [6]. So, for an accurate background, skin color evaluation color vision is adjudicated. Hence, Infrared and color both modalities are important for accurate detection. The features which get retrieved from the sensor data are used for making an Automatic Detection System (ADS). An ADS is basically made-up by the advancement of image processing and machine learning techniques.

Images play a vital role in human life whereas vision is probably the important sense of human beings. Machine learning combined as a significant part of computer vision when adaptation is required (e.g., human body recognition) because machine learning takes decision based on past experiences without human intervention [7].

The research work is taking advantage of both imageries (thermal and color) to improve the performance of victim detection with combination of the thermal-infrared spectra and color images with the help of machine learning. Results show that fusion is more accurate than individual modality.

2 Related Work

In recent years, an appraisal studies have completed that employ the infrared and color imaging in single and multi-camera configurations.

Basically, to find victims at catastrophe region, a set of features are extracted with a single camera. In color image model, features contain Gabor filter [8] or Haar wavelet [9] retorts, skin color, implicit shape models [10], image silhouettes [11], component-based gradient responses [12], and local accessible arenas [13]. In the same way, features are taken by monocular infrared technique. Such features include thermal hotspots [14], histograms of oriented gradients [15], shape-liberated histograms, contrast and inertial base features [16], body-model templates [17] and Scale Invariant feature transform. Human detection in thermal image contains some challenges like halos around cold or hot objects, lower smudging and resolution artifacts at the time of camera movement.

Some researchers have contributed using both thermal and color images such as Han et al. [18] proposed a hierarchical outline to analyze the initial human contour taken from infrared and color video. Aziz et al. [19] introduced a mobile vision modal to find out the victims by fusing thermal camera and video inputs. In his research work, a time effectual scanning technique was intended using a biologically stimulated search machine for the pan-tilt camera to expand the performance of the whole model. Pablo Tribaldos [20] proposed a technique to spotting humans in an insolent space by extracting the HOG features. This technique was applicable in both infrared and visible spectra. N Dalalet al. proposed a texture-based method [21] that applies Support Vector Machine (SVM) classifier using high-dimensional edges features to detect human body presence. Features are extracted from molecular images than applied to these features in SVM by various positive and negative samples. Some classification techniques in computer vision are distance matching [22], template matching [23], and convolutional neural networks [24] but most common classification technique is a Support Vector Machine (SVM) [25].

According to the above study, fusion of thermal and color imageries has ability to detect the presence of human body in all aspects with the incorporation of classification techniques.

3 Methodology

In this proposed work, fusion of color and infrared imaging is performed to improve the accuracy of victim detection. So, color images as well as infrared images are acquired in various environmental conditions by placing camera static at a position. After capturing images, analysis is performed in following scenario: Skin detection is applied on color images and local features descriptors are extracted from infrared images then features are passed on to a SVM classifier for detecting the presence of human. In this work (320×240) size images are used for prediction.

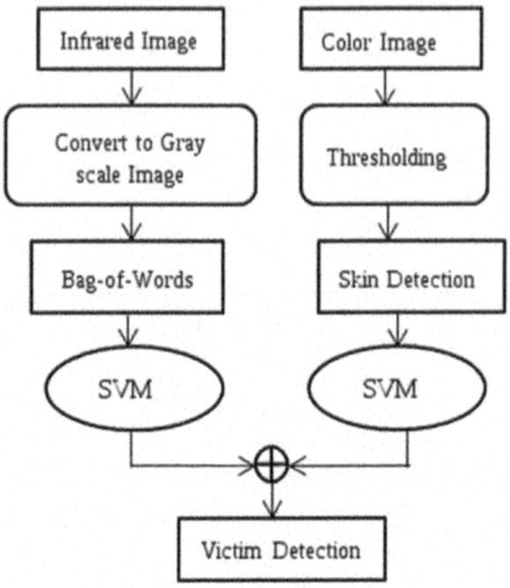

Fig. 1 Flow chart of the research work

3.1 *Infrared Image*

When Thermal image is a digital illustration of a scene and a proportion of the thermal radiation released by the objects. These images are taken by thermographic cameras, which can sense the radiation as infrared light [26]. Thermal imaging is a technique for enhancing perceptibility of objects in a shadowy situation by identifying the infrared radiation of objects and producing an image on the basis of that information.

In this research work (shown in Fig. 1), thermal images are transformed into gray-scale images because gray-scale images reduce the inherent complexity from a 3D pixel value to a 1D pixel value [27]. Basically, gray scale is the measurement of light intensity at each pixel as per the particular weighted combination of frequencies. After conversion, images are processed for victim detection by performing classification. Here Bag-of-Words model is applied for image classification by taking local features descriptor as words.

3.1.1 Bag-of-Words (BoW) Model

Bag-of-Words (BoW) model also termed as Bag-of-Features (BoF). In Computer Vision, BoW model treat a complete image as a document in which, "words" are defined in three steps: feature description, feature detection, and codebook generation. In this research work, three local feature descriptor such as Histogram of Gradient (HoG), Speeded up Robust Features (SURF), and Scale Invariant Feature Transform (SIFT) are applied to extract the features of an image. These features are basically described as histogram of their occurrence in the image [28]. After that, codebook is generated by a vector of identical words of a vocabulary of local features descriptors. In this work, three BoW clusters are used for analysis.

Then a SVM binary classifier is trained over these obtained histograms for training data corresponding to both positive and negative classes.

3.1.2 Local Feature Descriptors

In this work, three local feature descriptors well established for their performance in visible images have been explored for extracting and describing features in Infrared Images.

1. Histogram of Gradient (HoG) HoG is a method that provides a collection of image features in a schematic manner to analyze the objects enclosed in an image [29]. Later on, these features are used for object detection. In some cases, strong features are required for human detection. To obtain these features, first step is commenced to minimize the effect of lightning variation in image. After that histogram equalization is applied to normalize the color of an image. Generally, gamma normalization is used by $\sqrt{RGB}$ function.
   To compute Gradient computation: Smoothing value $\sigma = 0$ and filter kernels $Gx = [-101]$ and $Gy = [-101]^T$ are applied on x image that provides texture information, image contours, and shape.
   To generate the HOG descriptors, alignment binning splits the image into "cells" of $n \times n$ pixels.

   - Local normalization Dalal and Triggs [30] explain some various methods for image set normalization: L1-sqrt (Eq. 1), L1-norm (Eq. 2), L2-hys, and L2-norm (Eq. 3). In these equations, p is not a normalized vector and it contains each histogram in a given image set, ||p|| r, here r is 1, 2, e and r-norm are a constant. Finally, L2-hys and L2-norm are followed by clipping (v is 0.2) and re-normalizing. The normalization factor is one of the following:

$$L1\text{-}sqrt : f = \sqrt{p}(||p||1 + e) \tag{1}$$

$$L1\text{-}norm : f = p(||p||1 + e) \tag{2}$$

$$L2\text{-}norm : f = p\sqrt{(||p||22 + e2)} \tag{3}$$

   - HOG descriptors It is a final step. In this phase, processing of all blocks is performed and combined them in a compact grid of overlying chunks that covers the selection window to achieve the feature vector.

2. Scale Invariant Feature Transform (SIFT) This technique proposes local image features that are immutable to image translation, rotation, scaling, and partially immutable to illumination change and 3D projection [31], in the following scenario.

   - Scale-Space Extrema Detection It is represented as a function $F(a, b, \sigma)$ which is induced from an inconstant-scale Gaussian, $g(a, b, \sigma)$, using a target image, $I(a, b)$:

$$F(a, b, \sigma) = g(a, b, \sigma) * I(a, b) \tag{4}$$

   In Eq. 4, $*$ is representing the convolution operation between a and b,

$$g(a, b, \sigma) = \frac{1}{2\pi\sigma^2 e^{-(\frac{a^2+b^2}{2\sigma^2})}} \quad (5)$$

- Key point localization In this step, a comprehensive fitting of the closest data is applied for scale, ratio, and location of principal curves. This technique practices the Taylor expansion of scale-space function, $E(a, b, \sigma)$, to place the sample point at a origin.

$$E(\alpha) = E + \left(\frac{\delta E^T}{\delta a}\right)\alpha + \frac{1}{2}\alpha^T\left(\frac{\delta^2 E}{\delta a^2}\right)\alpha \quad (6)$$

  In above equation E and derivative of E are estimated at point and $(a, b, \sigma)T$ is the offset of this point.
- Key point descriptor Key point descriptors are dignified at a certain scale in the area of each key point. Basically, execution uses 4×4 descriptors from 16×16 that leads to a 128 $(4 \times 4 \times 8)$ element vector.

3. Speeded Up Robust Features (SURF) It is a robust local feature detector. Basically, it applies in computer vision for object recognition [32]. Somewhere, it enthused by the SIFT descriptor. So, SURF algorithm is used to extract feature point using scale space theory as similar to SIFT algorithm. In contrast, for feature point extraction, SIFT technique uses difference of Gaussians (DoG), whereas SURF technique adopts an integer estimate such as determinant of Hessian blob detector that predict rapidly with an integral image. For a particular pixel point of an image, its Hessian matrix can be denoted as M:

$$M(A, \sigma) = \begin{bmatrix} Caa(A, \sigma) & Cab(A, \sigma) \\ Cab(A, \sigma) & Cbb(A, \sigma) \end{bmatrix}, \quad (7)$$

   where C is a convolution of an image with the second derivative of Gaussian. According to the aim of increase the performance of SIFT algorithm, Gaussian filter is replaced by box filter. Furthermore, SIFT algorithm shortens the computation of determinant that not analyzes the weight of every region; therefore, the determining factor examine by the given equation:

$$Det(H) = \frac{\delta 2f}{\delta 2a2}\frac{\delta 2f}{\delta 2b2} - \frac{\delta 2f}{\delta 2a2\delta b} \quad (8)$$

   In above equation, $\frac{\delta f}{\delta a}$ is the convolution result of an image.

3.2 Color Image

Color images are also termed as RGB images. It contains additive color model that combines red, green, and blue light in variants ways to generate an extensive array

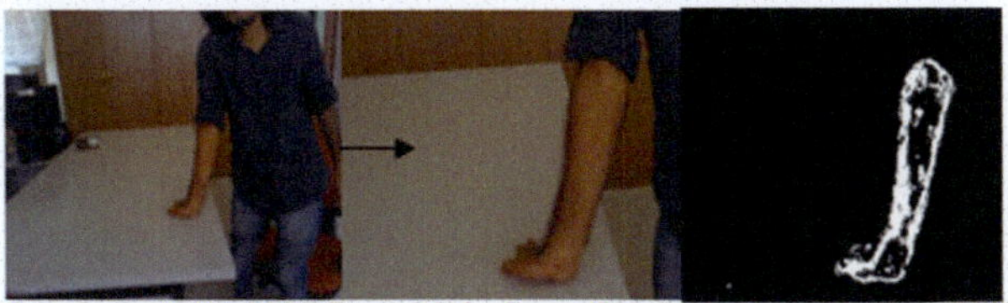

Fig. 2 Skin detection process

of colors. Color images generally stored as three separate raster maps in memory, where each map is treated as one channel. RGB color model is mainly used for detecting human body in daylight, warm environment, etc. In this proposed work, skin detection is performed using RGB color model.

3.2.1 Skin Detection Techniques

Skin is the key interface among the human body and its environment whereas skin color is a signal for detecting the presence of humans in images. Skin detection is basically detection of image pixels and area containing skin-tone color [33]. In this work, FLIR E60 camera is used that provides color images of resolution 2048 × 1536. So, color images are resized in 320 × 240 resolution from center then pre-trained non-parametric histogram-based model is used for differentiating skin and non-skin pixels. In this proposed work, skin detection is achieved in two stages: one is training stage and other is detection stage. Training stage follow these important steps:

1. Collection of images dataset that contains skin patches of different people under various lighting conditions.
2. Picking an appropriate space.
3. Consider the constraints of a skin classification model (generated by SVM).

Identification of skin pixels of a given image is performed by trained skin detection model as follows:

1. Color space of testing image should be same as color space of training phase.
2. Skin classifier is applied to classify each pixel in order to find that skin is present in a given image or not.
3. Spatial homogeneity on the detected regions is performed using morphology.

Figure 2 shows the complete process of skin detection. It represents the skin detection process from color images. It gives accurate result in the partial occlusion but fail to control full occlusion and miner presence of skin.

Table 1 Cardiovascular parameters

Proposed steps	Infrared images	Color images
Pre-processing	Gray scale images	Background modeling
Feature extraction	BoW (SIFT, SURF and HOG algorithm)	Skin color
Classification	SVM algorithm	SVM algorithm

3.3 *Support Vector Machine (SVM)*

It is a binary classifier technique that classifies the object by making hyperplane that separates the data according to class labels [34]. SVM technique performs on the basis of decision planes. A decision plane is used to separate the objects that have diverse class labels.

A hyperplane behaves like a decision plane to split up the set of input variable. In Support Vector Machine, a hyperplane is chosen to best divide the data points on the basis of their class, either 0 or 1. In two-dimensional space it is represented as a line and all data points are assumed to be separated by this line. For example

$$I0 + (C1 * V1) + (C2 * V2) = 0 \tag{9}$$

In Eq. 9, C1 and C2 are coefficients that are used to examine the slop of the hyperplane. Io represents the intercept that attained by learning technique and V1 and V2 are two data points. Classification is done by using "radial basis function kernel" (rbf) in SVM. Overall description of methodology is presented in Table 1.

4 Experimental Design

Data is acquired by FLIR E60 infrared camera which provides CSV file of generated infrared images of resolution 320 × 240. Camera provides color images of resolution 2048 × 1536. So, first color images need to resize in the 320 × 240 resolution then process for further evaluation. Optically registered IR and visible images of humans are captured during day and night. The data consisted of different pose and visibility variations of human such as human body completely visible, occluded human body, and only specific body part such as hand, leg, and head were visible. The data comprises two classes—positive and negative. Positive class containing human and negative class comprising of only background as shown in Fig. 3.

Table 2 shows the detection of human body in color and Infrared under various possible conditions. Table represented the advantages and limitation of both modalities.

Illustrative set of visible and IR images of negative class âĂŞ background

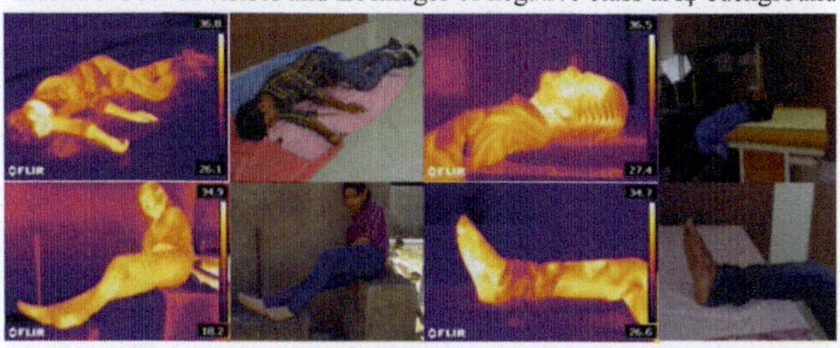

Illustrative set of visible and IR images of Positive class âĂŞ presence of victim

Fig. 3 The data comprises two classes in different environmental conditions

Table 2 Ease of manual detection of victim presence under different scenarios

Proposed steps	Infrared images	Color images
Daylight	High	High
Dark, sun light, fog and dust	BoW (SIFT, SURF and HOG algorithm)	Skin color
Warm environment (∼37 °C)	High	Low

5 Experimental Results

Experiments are performed in daylight and night. 450 images are used for daylight dataset in which 233 images are used for Bag-of-Words (BoW) model and 117 images are test images whereas 200 images are used as night dataset in which 50 images are used for BoW model and 80 images are used as test images. Dataset consists of two classes—humans as positive and all other objects as negative images. For algorithmic evaluation, the complete dataset is separated into three parts. First part for BoW model, second part to train the system, and third part to test the system. Performance of system is defined using test data which contains both positive and negative images in infrared imagery, color imagery, and Multi-modality fusion.

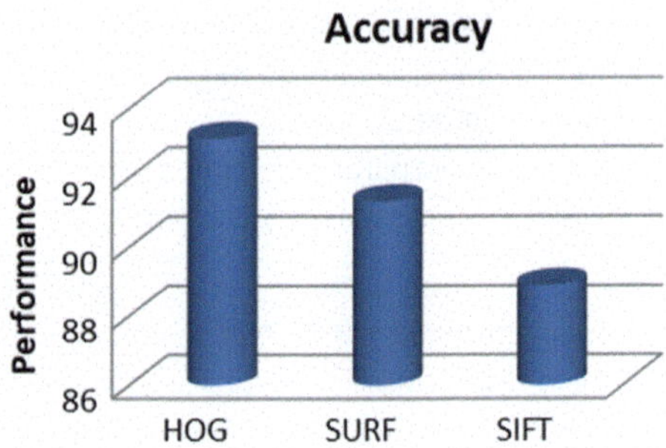

Fig. 4 Comparison on the basis of accuracy

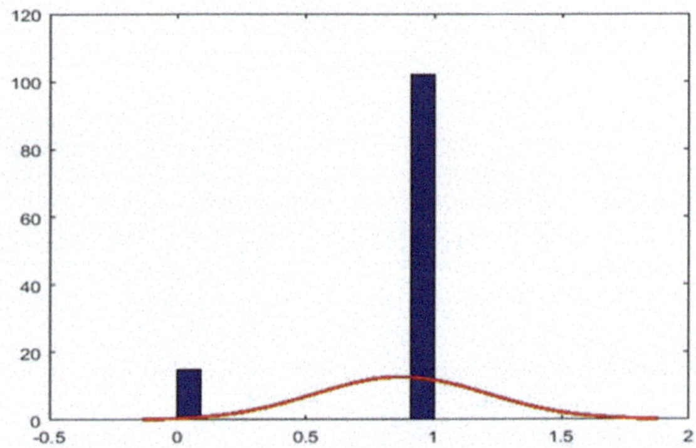

Fig. 5 Histogram of skin detection

5.1 *Infrared Imagery*

Victim detection classification model is performed on three local feature descriptors (HoG, SURF, and SIFT) extracted by infrared images. Figure 4 shows that Histogram of Gradient (HoG) outperforms SURF and SIFT techniques. Figure 4 shows the individual result of local feature descriptors in which HOG (93.12%) perform better in comparison to SURF (91.31%) and SIFT (88.89%). So, HoG is opted for multi-modality fusion.

5.2 *Color Imagery*

In color Imagery, skin color detection is performed to recognize the human body. In images, skin color is a sign of the presence of humans at a particular space. Here SVM is performed to categorize the human body on the basis of skin color. Results show that algorithm can be 80.34% accurately classify the data. In Fig. 5, one specifies the presence of skin and zero is for misdetection which is mainly due to full occlusion and minute presence of skin.

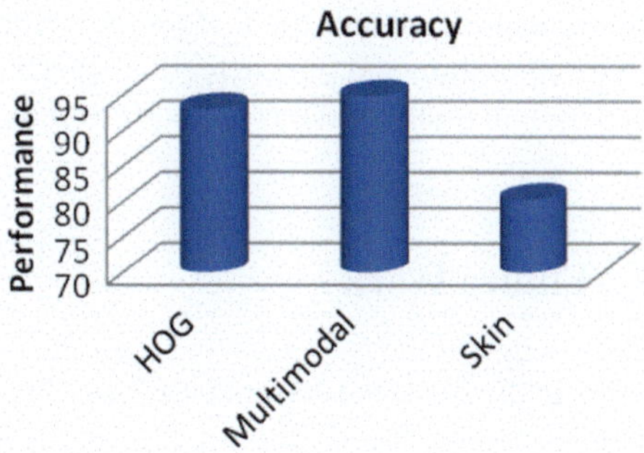

Fig. 6 Comparative analysis of infrared modality (HOG), color modality (Skin), and multimodality (IR + Color) on the basis of accuracy

Table 3 Performance measures of multimodal in daylight

Accuracy (%)	Precision (%)	Recall (%)	Environment
94.87	96.40	100	Daylight
94.12	95	98	Night

5.3 *Multimodality Fusion*

In this section, Infrared Imagery and color imagery are fused for victim detection. From infrared imagery, Hog features are opted because it outperforms other local descriptors and skin color is opted from color imagery. Here, Fig. 6 shows the comparison of Infrared modality (HoG), Color modality (Skin Detection), and Multimodality. In which Multi-modal performs better.

Table 3 represents the performance comparison of multi-modalities (HoG+Skin) individually in day time and night. The accuracy is dependent on the performance of sensors, efficient features like extraction, training, and classification.

6 Conclusions

In this research work multimodal imaging is applied to detect the human body in disaster environment. Infrared and color both visions have advantages in this problem domain. Color images are excellent to detect shape, color, and texture of objects in day time, whereas infrared images work equally well in day and night. Skin detection is applied on color imaging whereas Speeded up Robust Features (SURF), Scale invariant feature transform (SIFT), and Histogram of Gradient (HOG) methods are applied on infrared images for the improved detection of visible as well as occluded human body. Support Vector Machine (SVM) technique is used to classify the object. The proposed framework has learning-based automatic victim detection capability. Preliminary results indicate that victim detection was better in infrared images than skin color detection in visible images. In infrared Imaging, HOG algorithm performs better than SURF and SIFT algorithm, whereas in Multimodal Imagery, Fusion of HOG algorithm, and Skin detection algorithm performs better. So, for victim detection system, multimodal is more preferable.

7 Future Work

In the future work, the research work would be optimizing to improve the performance and processing speed. The application of more suitable classifier in the proposed system would be considered for more complex data. The real-time implementation of victim detection is also a scope for future.

Acknowledgements The scratch of this work is done in the lab of CSIR-CSIO, Chandigarh. I am also thankful to Mr. Satish Kumar, Sr. Principal Scientist, CSIR-CSIO, Chandigarh to sharing their precious wisdom during the course of this research.

References

1. B. Toft, S. Reynolds, *Learning from Disasters* (Springer, Berlin, 2016)
2. Disaster statistics (2018), http://www.worldfocus.in/magazine/disaster-management-in-india/
3. T.B. Bhondve, R. Satyanarayan, M. Mukhedkar, Mobile rescue robot for human body detection in rescue operation of disaster. Int. J. Adv. Res. Electr. Electron. Instrum. Eng. **3**(6), 9876–9882 (2014)
4. S.G. Narasimhan, S.K. Nayar, Shedding light on the weather, in *2003 IEEE Computer Society Conference on Computer Vision and Pattern Recognition, 2003. Proceedings*, vol. 1 (IEEE, 2003), pp. I-I
5. R. Usamentiaga, P. Venegas, J. Guerediaga, L. Vega, J. Molleda, F.G. Bulnes, Infrared thermography for temperature measurement and non-destructive testing. Sensors **14**(7), 12305–12348 (2014)
6. R. Gade, T.B. Moeslund, Thermal cameras and applications: a survey. Mach. Vis. Appl. **25**(1), 245–262 (2014)
7. O. Lézoray, C. Charrier, H. Cardot, S. Lefèvre, Machine learning in image processing (2008)
8. S. Zhou, Y. Jiang, J. Xi, J. Gong, G. Xiong, H. Chen, A novel lane detection based on geometrical model and Gabor filter, in *2010 IEEE Intelligent Vehicles Symposium* (IEEE, 2010), pp. 59–64
9. J. Krommweh, Tetrolet transform: a new adaptive Haar wavelet algorithm for sparse image representation. J. Visual Commun. Image Represent. **21**(4), 364–374 (2010)
10. T.F. Cootes, M.C. Ionita, C. Lindner, P. Sauer, Robust and accurate shape model fitting using random forest regression voting, in *European Conference on Computer Vision* (Springer, Berlin, Heidelberg, 2012), pp. 278–291
11. S. Milani, G. Calvagno, A depth image coder based on progressive silhouettes. IEEE Signal Process. Lett. **17**(8), 711–714 (2010)
12. D.A. Forsyth, O. Arikan, L. Ikemoto, J. O'Brien, D. Ramanan, Computational studies of human motion: part 1, tracking and motion synthesis. Found. Trends® Comput. Graph. Vis. **1**(2–3), 77–254 (2006)
13. C. Jepping, T. Luhmann, Object deformations from image silhouettes using a kinematic finite element beam model. Int. Arch. Photogramm. Remote Sens. Spat. Inf. Sci. 41 (2016)
14. S. Munder, D.M. Gavrila, An experimental study on pedestrian classification. IEEE Trans. Pattern Anal. Mach. Intell. **28**(11), 1863–1868 (2006)

15. F. Xu, X. Liu, K. Fujimura, Pedestrian detection and tracking with night vision. IEEE Trans. Intell. Transp. Syst. **6**(1), 63–71 (2005)
16. F. Tavakkoli, S. Ebrahimi, S. Wang, K. Vafai, Analysis of critical thermal issues in 3D integrated circuits. Int. J. Heat Mass Transf. **97**, 337–352 (2016)
17. Y. Fang, K. Yamada, Y. Ninomiya, B.K. Horn, I. Masaki, A shape-independent method for pedestrian detection with far-infrared images. IEEE Trans. Veh. Technol. **53**(6), 1679–1697 (2004)
18. A. Broggi, A. Fascioli, P. Grisleri, T. Graf, M. Meinecke, Model-based validation approaches and matching techniques for automotive vision based pedestrian detection, in *2005 IEEE Computer Society Conference on Computer Vision and Pattern Recognition (CVPR'05)-Workshops* (IEEE, 2005), pp. 1–1
19. M. Dantone, J. Gall, C. Leistner, L. Van Gool, Body parts dependent joint regressors for human pose estimation in still images. IEEE Trans. Pattern Anal. Mach. Intell. **36**(11), 2131–2143 (2014)
20. J. Han, B. Bhanu, Detecting moving humans using color and infrared video, in *Proceedings of IEEE International Conference on Multi-sensor Fusion and Integration for Intelligent Systems, MFI2003* (IEEE, 2003), pp. 228–233
21. M.Z. Aziz, B. Mertsching, Survivor search with autonomous UGVs using multimodal overt attention, in *2010 IEEE Safety Security and Rescue Robotics* (IEEE, 2010), pp. 1–6
22. C. Zhao, W. Shi, Y. Deng, A new Hausdorff distance for image matching. Pattern Recognit. Lett. **26**(5), 581–586 (2005)
23. T. Dekel, S. Oron, M. Rubinstein, S. Avidan, W.T. Freeman, Best-buddies similarity for robust template matching, in *Proceedings of the IEEE Conference on Computer Vision and Pattern Recognition* (2015), pp. 2021–2029
24. M. Szarvas, A. Yoshizawa, M. Yamamoto, J. Ogata, Pedestrian detection with convolutional neural networks, in *Intelligent Vehicles Symposium* (2005), pp. 224–229
25. P. Tribaldos, J. Serrano-Cuerda, M.T. López, A. Fernández-Caballero, R.J. López-Sastre, People detection in color and infrared video using HOG and linear SVM, in *International Work-Conference on the Interplay Between Natural and Artificial Computation* (Springer, Berlin, Heidelberg, 2013), pp. 179–189
26. L. Andreone, F. Bellotti, A. De Gloria, R. Lauletta, SVM-based pedestrian recognition on near-infrared images, in *ISPA 2005. Proceedings of the 4th International Symposium on Image and Signal Processing and Analysis* (IEEE, 2005), pp. 274–278
27. M. Vollmer, K.P. Möllmann, *Infrared Thermal Imaging: Fundamentals, Research and Applications* (Wiley, New York, 2017)
28. W. Burger, M.J. Burge, M.J. Burge, M.J. Burge, *Principles of Digital Image Processing* (Springer, London, 2009), p. 221
29. C.F. Tsai, Bag-of-words representation in image annotation: a review. ISRN Artif. Intell. (2012)
30. N. Dalal, B. Triggs, Histograms of oriented gradients for human detection (2005)
31. X. Yuan, J. Yu, Z. Qin, T. Wan, A SIFT-LBP image retrieval model based on bag of features, in *IEEE International Conference on Image Processing* (2011), pp. 1061–1064
32. H. Bay, T. Tuytelaars, L. Van Gool, SURF: speeded up robust features, in *European Conference on Computer Vision* (Springer, Berlin, Heidelberg, 2006), pp. 404–417
33. A. Elgammal, C. Muang, D. Hu, Skin detection, *Encyclopedia of Biometrics* (2015), pp. 1407–1414
34. C.C. Chang, C.J. Lin, LIBSVM: a library for support vector machines. ACM Trans. Intell. Syst. Technol. (TIST) **2**(3), 27 (2011)

A Novel Approach to Classify Breast Cancer Tumors Using Deep Learning Approach and Resulting Most Accurate Magnification Factor

Mukta Sharma, Rahul Verma, Ambuj Mishra, and Mahua Bhattacharya

Abstract In the recent research, breast cancer has come out to be the biggest reason behind death among females. Detection of breast cancer in its earlier stages is really the need of time. Detection of cancerous tumor is really a long and time taking process which is very costly and requires a lot of workforce as well. Therefore, providing a computer assisted approach for the simpler classification areas could be a solution. This could simplify the complex examine process of pathologists. Hence, the most appropriate magnification factor based CNN framework is proposed which leads to lesser expenses and lesser workforce. The proposed framework considers the different magnification factor images to obtain the classification accuracy. The proposed CNN-based framework is sectioned into three parts: preprocessing, feature extraction based on CNN, and CNN-based classification. The preprocessing phase consists of four processes, i.e., reshaping, formatting, image labeling, and train–test splitting. These preprocessings are applied after training and testing procedure of CNN model. In experimental analysis, publicly available breast cancer dataset from histopathology is used. The dataset consists of seven thousand nine hundred and nine images obtained from 82 different patients. These images are distributed among four different magnification factors: 40X, 100X, 200X, and 400X. These magnification factors are utilized for training of CNN model and for obtaining the accuracy. In result

M. Sharma (✉) · R. Verma · A. Mishra · M. Bhattacharya
ABV-Indian Institute of Information Technology and Management,
Morena Link Road, Near IDBI Bank ATM, Gwalior, India
e-mail: mukta.24sharma@gmail.com

R. Verma
e-mail: rahulverma.upe@gmail.com

A. Mishra
e-mail: ambuj.mishra.2k12@gmail.com

M. Bhattacharya
e-mail: mahuabhatta@gmail.com

A. Nanda and N. Chaurasia (eds.), *High Performance Vision Intelligence*,
Studies in Computational Intelligence 913,
https://doi.org/10.1007/978-981-15-6844-2_13

analysis, four different experimentations are conducted to show the effectualness of the presented CNN-based framework. Also, the results exhibit that the CNN-based framework used in this work achieves the highest average accuracy of 97.63% among the other architectures and existing approaches.

1 Introduction

Breast cancer categorical classification of histopathological images utilizing the deep neural system is noteworthy for clinical determination and forecast with the dispatch of the exactness medication activity. As indicated by the report from the WHO, breast cancer is the main well-known disease with high grimness and mortality among ladies worldwide. Malignant growth is a basic general therapeutic issue on mankind today. According to WHO, the world has just seen 8.2 million passing as a result of breast cancer in 2012 and these numbers could reach up to 27 million preceding 2030. Particularly, BC is the major reason of deaths amid women. Death percentage due to BC is quite high in comparison to other forms of cancer. According to Cancer Country Profiles released in 2015 by WHO, 25% out of 326.3K women in India died because of BC, average life expectancy at birth for females was found to be 68. Age-standardized cancer mortality was found to be 15 year in 2007 which further decreased to 13 years in 2013. Presently, manual BC supervised learning for cytological images is a hard errand. Breast cancer histopathological imaging is really a time taking process which gets affected by drop in attention and fatigue and is highly dependable on the expertise of the pathologist. According to [11], a dire need is required for one computer-based diagnosis just to decrease the pressure from doctors by easily classifying the benign portion from malignant portion. Therefore, there is a need to concentrate on areas of the cell images that require higher expertise to diagnose [9]. Thus, a lot of time and work has been invested by the researchers in analysis of histopathological images of BC, especially to design an automated diagnostic system to classify benign and malignant cells. There are various approaches that have been developed for the pathological images starting from rule based to classical machine learning applications [20]. In the recent era, the deep learning is utilized in various image analysis procedures which outperforms the conventional approaches [7, 13, 14]. The deep CNN successfully solves the problems of medical imaging domain by showing their potential in diagnosis of breast cancer [1, 3, 20, 25]. Kowal et al. [15] have performed different algorithms on a deadest of 500 images to perform nuclei segmentation and compared their results. There is a critical need of a programmed classifier for arranging favorable pictures with harmful pictures. Therefore, to fulfill the requirement CAD has created noteworthy measure of work in a similar field. Some of them are discussed below.

Wang et al. [27] proposed a novel deep convolution network based approach to detect metastases cancer in complete slide images of breast sentinel lymph node. The authors worked on patch level to differentiate the cancerous patches from benign patches. They have trained the deep CNN with millions of both types of patches.

They isolate the predictions related to patch level and use these predictions to plot tumor probability heatmaps. Thereafter, they carried out posterior process on these heatmaps to evaluate prognostication for the slide-based classification and to localize the tumor. Finally, they merge their predictions with the expert pathologists' interpretations which help them to reduce the manual error upto a significant rate. Similarly, Cireşan et al. [8] proposed a DNN-based approach as a powerful pixel classifier. The proposed approach utilized the raw RGB pixels sampled from square patches of source images. The approach classified the patch containing mitotic nucleus near the center from other patches by learning visual features during training using a classifier. The proposed approach contributes in the area of classification, segmentation, and detection and outperforms the other existing approaches. Rakhlin et al. [19] introduced a novel computational approach based on deep CNN and gradient boosted tree classifiers to separate the histopathology pictures. They have utilized the potential of deep CNN and gradient boosted tree for extraction of feature and classification, respectively. The results are evaluated using the ICIAR 2018 Breast Cancer Images [2]. The approach achieved the 87.2% accuracy for four class classification and 93.8% for two class classification. The results show that the proposed approach performs far better in comparison to the similar approaches.

Spanhol et al. [24] have shown in the paper the exhibition of most regular order frameworks depending on suitable information portrayal. And a significant part of the endeavor is devoted to featuring engineering, a troublesome and tedious procedure. It utilizes earlier master area learning of the information to make helpful highlights. Thereafter, profound learning can remove and sort out the segregated data from the information, not requiring the structure of highlight extractors by an area master. Wong et al. [28] explored the advantage of expanding information with artificially made examples while preparing an AI classifier. The authors tentatively assess the advantages of information growth for a convolution backpropagation-prepared neural system. Krizhevsky et al. [16] prepared an extensive, profound convolutional neural system to order the 1.3 million huge-goal images in the LSVRC-2010. The proposed CNN model named as ImageNet was ready to feed by the 1000 distinct classes. For the test data, the authors accomplished top-1 and top-5 blunder rates of 40% and 19% which are impressively very high to the previously achieved class results. CNN model consists of 60 millions parameters and 500,000 neurons, having five convolutional layers along max-pooling layers, and two softmax layers as final layers. Hamilton et al. [12] show the genomic transformation which has prompted quick development in the successiveness of qualities and proteins, and consideration is currently swinging for the capacity of the conceal proteins. This enables protein restriction for the images with increased throughput. Therefore, it is required for extensive mount robotized computing methods to proficiently evaluate, recognize, and order sub-cell images.

Litjens et al. [18] raise the issues that the pathologists had faced. Due to customized prescription a considerable increment has been done in remaining task at hand and multifaceted nature of histopathological malignancy conclusion. Along these lines, symptomatic conventions need to concentrate similarly on proficiency and precision. This paper presents "profound learning" as a system to improve the objectivity and proficiency of histopathological slide examination. Through two precedents, prostate malignant growth is distinguished from proof in biopsy examples and bosom disease metastasis location in sentinel lymph hubs. Chen et al. [6] proposed a novel work to measure the quantity of mitoses per tissue zone. This gives a significant forcefulness sign of the intrusive bosom carcinoma. However, programmed mitosis identification in histology data remains a difficult issue. Customary strategies either utilize hand-created highlights to separate mitoses from different cells or build a pixel-wise classifier to mark each pixel in a sliding window way. While, based on previous experiences the expansive shape variety of mitoses and the presence of numerous mirrors with comparative appearance, the moderate speed of the later restricts.

There are a few application areas that utilize common element rubrics on double division for microscopic pictures of breast cancer. Spanhol et al. [24] used a BC histopathology dataset (BreaKHis), and at that point gave a pattern of binary classification. For a given histopathological image, CNN will extract the features present in the images. However, deep learning approaches do not need a definite structure for feature extraction and select the distinguishable information from the data to use it further for the classification process at the last layers of the CNN [21]. Selecting feature during conventional machine learning methods takes more time and reduces the efficiency of the model. The authors have demonstrated that the CNN-based profound learning structure indicates preferable arrangement execution over the hand-crafted methods [22]. After having numerous number of existing approaches for breast cancer classification, segmentation, and detection, still the need of highly automated and accurate computerized diagnosis system is not accomplished. Therefore, a highly automated and accurate CNN-model-based framework is required for this classification task. Mainly, two factors are considered in this research for all the images: area covered and clarity of the images, since lower magnification factors cover larger piece of area than higher magnification factors. Thus, considering area covered as dominating factor, lower magnification factors should produce higher classification accuracy. However, higher magnification factors will produce more clear images than lower magnification factor. Therefore, considering clarity as dominating factor, higher magnification factors should produce better classification accuracy.

Thus, in this work hyperparameters are tuned and a novel CNN-based method is given for classification of breast cancer cells by using four different magnification factors. The proposed framework is categorized into three phases: (i) preprocessing, (ii) feature extraction, and (iii) classification. In preprocessing phase, whole dataset images are gone through the four processes, i.e., reshaping, formatting, image labeling, and train–test splitting. After preprocessing, the hyperparameter-tuned CNN model is trained using these preprocessed images of different magnification factors based on feature extraction and classification phases of the CNN architecture. After training, the CNN model is tested by applying the images of different magnification factors of breast cancer cell dataset [24] and the accuracy for each magnification factor images is obtained. In this work, four types of experimentations are conducted to illustrate the effectiveness and strength of the presented framework. Further, other sections of the paper are formulated as follows: Sect. 2 presents the complete description for the proposed CNN-based framework. Section 3 provides the validation aspects of the proposed framework. Lastly, Sect. 4 concludes the overall proposed CNN-based framework and gives the future aspects.

2 Proposed Methodology

Classification of tumor is quite complex problem and contains multiple steps in between. All of these steps have their own significance and can affect the final results in multiple ways possible. Our work is to provide a CAD after the medical imaging process to reduce the workload of pathologists by classifying the easily detectable images. The schematic diagram of the overall proposed approach is illustrated in Fig. 1. For CNNs, selecting a suitable network architecture is complicated. The overall deep CNN-based proposed architecture is inspired by the existing CNN model GoogleNet [26]. Specifically the architecture consists of six layers as shown in Fig. 2. Initially, there are four convolution layers alternatively merged with pooling layers. Remaining two layers are fully connected layers employing classification. The overall proposed CNN-based framework is categorized in three phases as shown in Fig. 2 as follows:

1. Preprocessing,
2. Feature extraction, and
3. Classification.

The proposed CNN-based framework is comprised of following components: network architecture, preprocessing, model training, feature extraction, and classification.

2.1 Network Architecture

1. **Convolutional Layer**: It is assumed that this layer is the Kth layer of the network. At this layer, N^K represents the feature map and K is considered as superscript. Therefore, every feature map is represented as $M_x^K (j = 1, 2, 3, \ldots, N^K)$. F_{ix}^K is an array of 2D filters used to parametrize this convolutional layer. It also associates the pth feature map M_p^{K-1} in the $(K-1)^{th}$ layer with the xth feature map M_x^K in the Kth layer and the bias b_x. Here every filter pretends like a detector which is capable of detecting feature of one particular type. This is achieved by convolving all the locations of feature map. The feature map $M_i^{K-1} (i = 1, 2, \ldots, N^{K-1})$ is convolved with F_{ix}^K to obtain F_x^K. The results obtained are added and appended along the bias b_x. Now in element-wise order, the non-linear activation function $\phi(.)$ is attained. The equation given below represents the feature map of Kth layer:

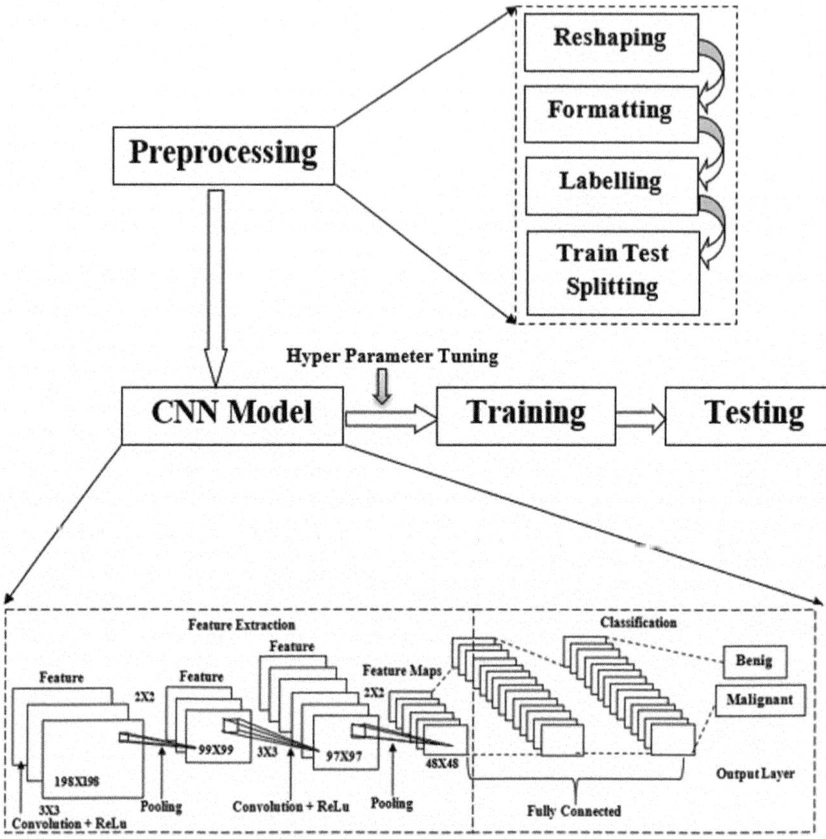

Fig. 1 Schematic diagram of overall proposed approach

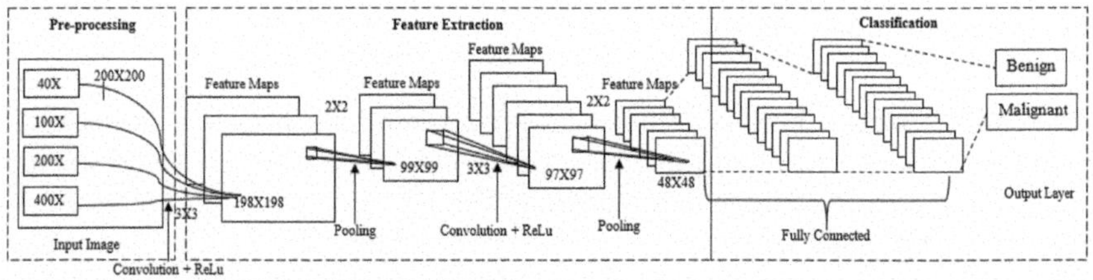

Fig. 2 Architecture of proposed CNN model

$$M_x^K = \phi\left(\sum_{i=1}^{N^{K-1}} M_i^{K-1} * F_{ix}^K + b_x^K\right), x = 1, 2, \ldots, N^K \tag{1}$$

where * represents the convolution operator.

2. **Pooling Layer**: A pooling layer is used for dimensionality reduction of a feature map. Utilization of this layer aids in reduction of the complexity and time to train the CNN models and also adds non-uniformity in small translations of input images. Two most commonly applied methods are *max-pooling* and *average pooling*. First one chooses the extreme activation on top of a compact pooling region. However, the other one evaluates the mean activation on top of this region. Max-pooling is generally superior in performance over the average pooling [5].
3. **Classification Layer**: Generally, fully connected layers are the top layers of classification layer in CNN architecture. Since the proposed framework consists of two layers named as fully connected layers where it (F_6 in Fig. 2) takes the plunged feature maps of 5th layer as input. The final fully connected layer fulfills the end product purpose, i.e., predictions. The second last layer F_6 output (represented as M^6) is provided as input to final layer F_7 which comprises weights W_7 and biases b_7. To relate with the n classes of staining patterns, F_7 layer has n number of neurons. The predictions $\hat{q} = [\hat{q}_1, \hat{q}_2]^T \in R_n$ via softmax regression are as follows:

$$M^7 = W^7 M^6 + b^7, M^7 \in R_n \tag{2}$$

$$\hat{q}_x = \frac{exp(M_x^7)}{\sum_{i=1}^{n} exp(M_i^7)} \tag{3}$$

where $\hat{q}_x$ is the end layer predictions of the xth neuron.

2.2 Preprocessing

A suitable image preprocessing procedure is required for deep CNNs that intake the appropriate features of images for proper examination and release of optimal local feature representation and classification outcome. The dataset taken into consideration has poor contrast and cell overlapping issues. To overcome these issues, the images are preprocessed as follows:

1. Reshaping,
2. Formatting,
3. Image labeling, and
4. Train–test splitting.

Reshaping: All the images of the dataset are very large in size, which makes the accessing of images difficult and consumes large memory. Therefore, all the images of dataset belonging to Benign and Malignant categories are reshaped to a specific dimension (200X200 in our work).

Formatting: Images are captured in four different magnification sizes (40, 100, 200, 400), and they are divided into different categories based on the cancer type. Dataset provided is available in a very complex way with multiple folders and sub-folders present in it which makes the implementation very hard. So, the dataset has been converted into another format so that it can be provided as an input into the model. Now, the modified dataset is divided into different folders based on the cancer type.

Labeling: Images are in a single folder that contains all types of cancer images. Before training the model labeled images are needed. So, we have to iterate over all the files in that folder and then split its path and sort them into a Python numpy array. Python package *Shutil* and *os* is used for the image labeling with the help of Python list.

Train–test splitting: The data is generally partitioned for the testing and training of the model. In the training section, images and their labels are known beforehand and the model is trained. Here, the presented work splitted the training and the testing data into three parts, i.e., 90%(Train)-10%(Test), 80%(Train)-20%(Test), and 70%(Train)-30%(Test).

2.3 CNN Model Training

Since the deep CNNs require network coverage to fast optimal solution (because of its non-convex property of the defined cost function), suitable parameter settings are necessary. The defined weights (W^p) and the biases (b^p) of the different convolution layers as well as fully connected layers are used to parametrize the deep CNN model. For $W^p \& b^p$, $p = 1, 3, 5, \ldots$. Here, it is expected that the total number of parameters that can be trained is greater than 60,000. The model is trained by minimizing the log loss (cost function) among the output probability vector $\hat{q} = [\hat{q}_1, \hat{q}_2]^T$ and the binary label vector $q = [q_1, q_2]^T$ which has single non-zero entry "1" representing the true class.

$$C_F(q, \hat{q}) = -(qlog(p) + (1 - q)log(1 - p)) \tag{4}$$

where p is the model's predicted probability. Initialization of weights is done using uniform distribution between $(-r, r)$ with $r = \sqrt{\frac{6}{fan\text{-}in + fan\text{-}out}}$ [29], where fan-in and fan-out are the number of inputs and outputs to each neuron.

2.4 Feature Extraction and Classification

Preprocessing as described in Sect. 2.2 is performed on images before the feature extraction task for training and testing purpose. There are four layers (two convolution layers alternated with two pooling layers) used for feature extraction in the proposed framework. The processed image is then promulgated through the network to the next layer (C_1 in Fig. 2). In this way, the feature map is obtained in each layer to evaluate the probability of the cell for each class. The estimated class is the one maximum among the two probabilities.

3 Experimental Evaluations and Discussions

3.1 Dataset

The validity of the proposed CNN-based framework is done by analyzing its performance using standard publicly available BreakHis dataset [24] which is used. The dataset is based on breast cancer cell images acquired by microscope, utilizing various magnification factors (the enlargement procedure for images visible by microscopic lens known as magnification factors). There are 7909 images which were taken from 82 patients which are divided into 2 categories: benign and malignant. For the framework proposed in this chapter, the various magnified images (40X, 100X, 200X, and 400X) were utilized for evaluating the classification accuracy. To execute the proposed framework, 80% of the total dataset is used to train while 20% of the dataset is used to validate and evaluate the accuracy of the proposed architecture for each magnification factor. Furthermore, the basic stopping criteria (1000 iterations in this case) is delineated based on the performance validation to overcome the overfitting issues.

3.2 Results and Analysis

There are four different experiments that have been performed to analyze the performance of the proposed framework. In experiment-1, the CNN architecture is executed using the preprocessed images of the dataset. Then the CNN model is trained using 80% of the dataset. Thereafter, the CNN architecture is then made to interpret 20% of the dataset images. The architecture of the CNN model is shown in Table 1. The architecture consists of two convolution layers, two pooling layers, and two fully connected layers. The accuracy achieved by the architecture was only 75.4%. Thus, the experiment-2 is then performed after tuning the hyperparameters of the CNN architecture shown in Table 2. Designing and training of the deep CNN models is the crucial task and requires making many choices. For experiment-2, tuning of hyper-

Table 1 Architecture of the CNN model before tuning of hyperparameters

Layer (Type)	Output shape	Param #
conv2d_1 (Conv2D)	(None, 148, 148, 32)	896
activation_1 (Activation)	(None, 148, 148, 32)	0
max_pooling2d_1 (Maxpooling2)	(None, 74, 74, 32)	0
conv2d_2 (Conv2D)	(None, 72, 72, 32)	9248
activation_2 (Activation)	(None, 72, 72, 32)	0
max_pooling2d_2 (Maxpooling2)	(None, 36, 36, 32)	0
conv2d_3 (Conv2D)	(None, 34, 34, 64)	18496
activation_3 (Activation)	(None, 34, 34, 64)	0
max_pooling2d_3 (Maxpooling2)	(None, 17, 17, 64)	0
flatten_1 (Flatten)	(None, 18496)	0
dense_1 (Dense)	(None, 64)	1183808
activation_4 (Activation)	(None, 64)	0
dropout_1 (Dropout)	(None, 64)	0
dense_2 (Dense)	(None, 1)	65
activation_5 (Activation)	(None, 1)	0

parameters is done by using the guidance given in [4]. The verification of the trained CNN architecture is done by using the hit-n-trial method. After various hyperparameter setting investigations, the CNN architecture is designed as shown in Table 3 and the total number of non-trainable parameters detected is 576, number of epochs is 50, and number of hidden layers is 4. When talking about the most suitable activation function, then non-linear activation functions are the most preferable ones for the neural networks. Due to their capability of providing the network nodes to understand more complicated structures in the data, the two most popular conventional non-linear activation functions are the *sigmoid* and *hyperbolic tangent* activation functions. It is observed that both these activation functions have the common vanishing gradient issue [10], which prevents the deep CNN from learning effectively. The *ReLu* activation function very well handles the vanishing gradient problem and allows the deep CNNs to learn faster and achieve better performance. The ReLu function looks and acts like a linear activation function. However, in real it is a non-linear function which allows the nodes to understand more complicated structures in the data. Therefore, in this work the proposed CNN architecture adopted the ReLu function as an activation function.

The results of proposed framework obtained from the experiment-2 are compared with the other existing architectures. These are GoogleNet [26], VGGNet [23], and ResNet [29]. As shown in Table 4, the average classification accuracy (ACA) based on four magnification factors given by the GoogleNet, VGGNet, and ResNet individually are 93.75%, 93.4%, and 93.97%, respectively. However, the proposed framework gives the ACA of 97.37%. This can be observed from these results that the proposed framework gives the better performance in terms of ACA in comparison to other

Table 2 Hyperparameters obtained relevant to CNN model

Layer	Layer type	Hyperparameters	Computational complexity
Input	Input	Image Size wxw: 200x200	$O(w^2)$
C_1	Convolution	Filter size $k_1 X k_1$: 32X32 Activation function Relu, y = max(0, x)	$O(w_n^2, k_1^2)$
P2	Pooling	$k_2 X k_2$: 2X2 method: maxPooling	$O(w^2, n_1)$
C3	Convolution	$k_3 X k_3$: 3X3 Activation Function Relu, y = max(0, x)	$O(w^2, n_1)$
P4	Pooling	$k_4 X k_4$: 2X2 method: maxPooling	$O(w^2, n_2^2/k_2)$
F6	Fully connected	Neuron no. : $n_4 = 512$	$O(w^2, n_2^2/k_2)$
F7	Fully connected	Neuron no. : $n_5 = 128$	$O(w^2, n_2^2/k_2)$
Output	Output	Neuron no. : 2	$O(w^2, n_2^2/k_2)$

Table 3 Architecture of the CNN model after tuning of hyperparameters

Layer (Type)	Output shape	# of param
conv2d_1 (Conv2D)	(None, 148, 148, 32)	320
max_pooling2d_1 (Maxpooling2)	(None, 74, 74, 32)	0
batch_normalization_1 (Batch)	(None, 74, 74, 32)	128
conv2d_2 (Conv2D)	(None, 72, 72, 64)	18496
max_pooling2d_2 (Maxpooling2)	(None, 36, 36, 64)	0
batch_normalization_2 (Batch)	(None, 36, 36, 64)	256
conv2d_3 (Conv2D)	(None, 34, 34, 64)	36928
max_pooling2d_3 (Maxpooling2)	(None, 17, 17, 64)	0
batch_normalization_3 (Batch)	(None, 17, 17, 64)	256
conv2d_4 (Conv2D)	(None, 15, 15, 96)	55392
max_pooling2d_4 (Maxpooling2)	(None, 7, 7, 96)	0
batch_normalization_4 (Batch)	(None, 7, 7, 96)	384
conv2d_5 (Conv2D)	(None, 5, 5, 32)	27680
max_pooling2d_5 (Maxpooling2)	(None, 2, 2, 32)	0
batch_normalization_4 (Batch)	(None, 2, 2, 32)	1286
dropout_1 (Dropout)	(None, 2, 2, 32)	0
flatten_1 (Flatten)	(None, 128)	0
dense_1 (Dense)	(None, 128)	16512
dropout_2 (Dropout)	(None, 128)	0
dense_2 (Dense)	(None, 8)	1032

Table 4 Proposed CNN-based framework comparison with other CNN architectures using different magnification factors

CNN architectures	Magnification factors				ACA
	40X (%)	100X (%)	200X (%)	400X (%)	
GoogleNet	93.7	90.4	95.3	95.6	93.75
ResNet	93.2	91.5	95.4	93.5	93.4
VGGNet	94.6	90.8	95.7	94.8	93.97
Proposed framework	97.4	98.6	97.7	96.8	97.625

three architectures. And it can also be concluded from Table 4 that the most accurate magnification factor found is 100X because of the highest ACA achieved by the proposed framework as well as by the other three architectures for this factor 98.7%, 95.3%, 95.4%, and 95.7%, respectively.

In experiment-3, the dataset is split into three different variations of train–test data sub-categories. These are D_1:90%-10%, D_2:80%-20%, and D_3:70%-30%. 90%-10% train–test splitting means 90% of the whole dataset consumed for training while rest of the 10% data consumed for testing. In Table 5, the proposed framework compared with other three CNN architectures based on different train–test splitting of dataset is shown. In Table 5, the "Class Category" represents the cancer category (N or C), where "N" belongs to benign tumor and "C" belongs to malignant tumor. The performance metrics used to compare the performance of the proposed framework with the other three architectures are precision, recall, F1-score, accuracy, and ACA. It is found that the proposed framework proves to be the most accurate approach for breast cancer classification among all.

In experiment-4, the log loss is considered to be the cost function for the CNN-based proposed framework. To strengthen the conclusion drawn from the results given in Table 4 that the 100X magnification factor is the most accurate among the four magnification factors one more metrics evaluation is performed. The log loss parameter is evaluated for different values of epochs using proposed framework as shown in Fig. 3. In Fig. 3a–d, the cost function values are evaluated on different epochs for each magnification factor, i.e., 40X, 100X, 200X, and 400X, respectively. As it can be observed from Fig. 3a, c, d, i.e., for 40X, 200X, and 400X, the validation loss is not decreasing with the training loss. However, for 100X graph (from Fig. 3b), the validation loss is decreasing with the training loss. Thus, it can be concluded that the value of loss is minimum for the 100X factor in comparison to other magnification factor values. Hence, 100X magnification or 10X objective lens or 0.2 μm pixel size is the most accurate factor found in comparison to other magnification factors.

Table 5 Proposed CNN-based framework comparison with other CNN architectures using train–test splitting

CNN Arch.	Training–testing Data splitting	Class category	Precision	Recall	F1-score	Accuracy (%)	Average accuracy (%)
GoogleNet	D_1	N	0.91	0.93	0.92	93.54	93.11
		C	0.95	0.93	0.92		
	D_2	N	0.92	0.94	0.92	93.21	
		C	0.92	0.93	0.92		
	D_3	N	0.95	0.91	0.92	93.67	
		C	0.91	0.97	0.94		
VGGNet	D_1	N	0.89	0.94	0.93	93.23	94.34
		C	0.87	0.92	0.93		
	D_2	N	0.96	0.95	0.94	95.48	
		C	0.93	0.92	0.91		
	D_3	N	0.90	0.91	0.93	92.21	
		C	0.88	0.94	0.95		
ResNet	D_1	N	0.96	0.91	0.92	92.00	93.89
		C	0.92	0.94	0.93		
	D_2	N	0.92	0.86	0.91	94.68	
		C	0.87	0.92	0.91		
	D_3	N	0.89	0.91	0.88	90.32	
		C	0.90	0.92	0.93		
Proposed framework	D_1	N	0.95	0.96	0.98	97.65	96.67
		C	0.95	0.96	0.97		
	D_2	N	0.96	0.98	0.97	97.55	
		C	0.96	0.97	0.98		
	D_3	N	0.98	0.97	0.98	98.33	
		C	0.96	0.97	0.98		

3.3 *Proposed Approach Evaluation in Comparison with State of the Art*

Similarly, a comparative analysis is performed using the proposed framework (PF) to show the superiority over the five existing approaches as shown in Fig. 4. It can be noted from Fig. 4 that the well-known approaches [1, 8, 17, 19, 27] obtained the accuracy of 78.2%, 93.80%, 88.89%, 83.3%, and 96.28%, respectively, whereas the proposed framework obtained an accuracy of 97.63%. Thus, it can be concluded that the proposed framework outperforms the existing well-known approaches (Fig. 3).

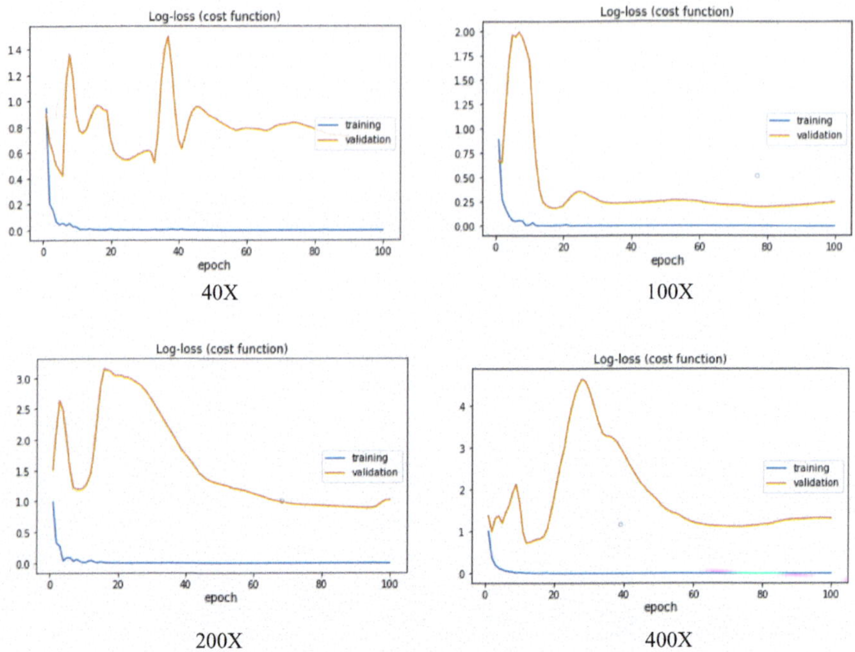

Fig. 3 Log loss function values during training and validation for each magnification factor

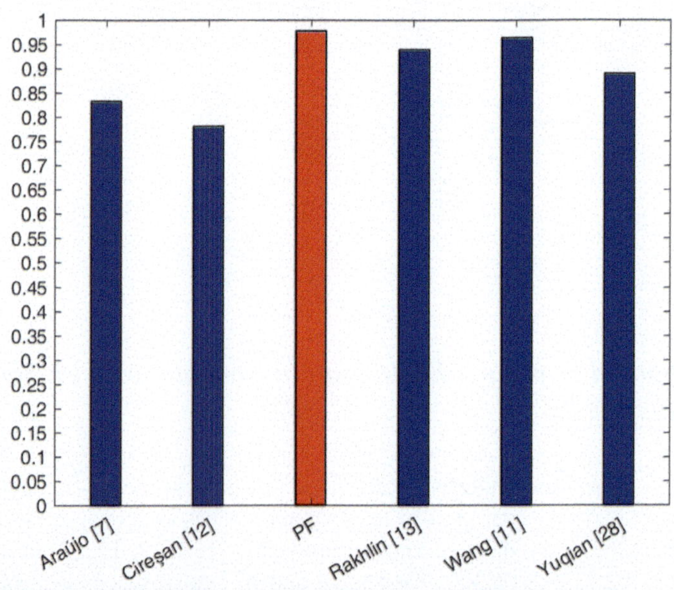

Fig. 4 Comparative analysis of proposed framework with the existing approaches

4 Conclusion

This paper proposed a novel deep CNN-based framework to classify the breast cancer cells using most accurate magnification factor images. The proposed framework comprised of three phases: (i) preprocessing, (ii) feature extraction, and (iii) classification. Preprocessing consists of four processes, i.e., reshaping, formatting, image labeling, and train–test splitting. In feature extraction process, the four layers of convolution layer with alternated pooling layers are used in CNN architecture. The feature maps are obtained from each layer output and given to classification phase of CNN architecture. Then, the two fully connected layers are used to classify the breast cancer cells using the feature maps as input from the previous layers. Finally, the proposed CNN-based framework performance is compared with other architectures based on accuracy for four different magnification factor images. From the experimental analysis, the proposed framework concludes that 100X is the most accurate magnification factors among the four. Similarly, the proposed framework is also compared with the existing approaches and the results show that it outperforms the existing approaches. It has been concluded from the overall result analysis that the paper gives contribution in the following manner:

1. Proposed a novel CNN-based framework to classify the breast cancer cell images.
2. Find the most accurate magnification factor of the BreakHis dataset images for future analysis.
3. Provide hyperparameter-tuned novel CNN architecture.
4. Achieves highest average accuracy of 97.63%.

The classification accuracy can be further improved by utilizing both hand-crafted features combined with CNN features. Different data augmentation techniques can also be used in future to improve the end output of the CNN architecture.

References

1. T. Araújo, G. Aresta, E. Castro, J. Rouco, P. Aguiar, C. Eloy, A. Polónia, A. Campilho, Classification of breast cancer histology images using convolutional neural networks. PloS One **12**(6), e0177544 (2017)
2. G. Aresta, T. Araújo, S. Kwok, S.S. Chennamsetty, M. Safwan, V. Alex, B. Marami, M. Prastawa, M. Chan, M. Donovan et al., BACH: grand challenge on breast cancer histology images. Med. Image Anal. **56**, 122–139 (2019)
3. B.E. Bejnordi, M. Veta, P.J. Van Diest, B. Van Ginneken, N. Karssemeijer, G. Litjens, J.A. Van Der Laak, M. Hermsen, Q.F. Manson, M. Balkenhol et al., Diagnostic assessment of deep learning algorithms for detection of lymph node metastases in women with breast cancer. JAMA **318**(22), 2199–2210 (2017)
4. Y. Bengio, Practical recommendations for gradient-based training of deep architectures, *Neural Networks: Tricks of the Trade* (Springer, Berlin, 2012), pp. 437–478
5. Y.-L. Boureau, J. Ponce, Y. LeCun, A theoretical analysis of feature pooling in visual recognition, in *Proceedings of the 27th International Conference on Machine Learning (ICML-10)* (2010), pp. 111–118

6. H. Chen, Q. Dou, X. Wang, J. Qin, P.A. Heng, Mitosis detection in breast cancer histology images via deep cascaded networks, in *30th AAAI Conference on Artificial Intelligence* (2016)
7. T. Ching, D.S. Himmelstein, B.K. Beaulieu-Jones, A.A. Kalinin, B.T. Do, G.P. Way, E. Ferrero, P.-M. Agapow, W. Xie, G.L. Rosen et al., Opportunities and obstacles for deep learning in biology and medicine. biorxiv (2017)
8. D.C. Cireşan, A. Giusti, L.M. Gambardella, J. Schmidhuber, Mitosis detection in breast cancer histology images with deep neural networks, in *International Conference on Medical Image Computing and Computer-Assisted Intervention* (Springer, 2013), pp. 411–418
9. C. Désir, C. Petitjean, L. Heutte, M. Salaun, L. Thiberville, Classification of endomicroscopic images of the lung based on random subwindows and extra-trees. IEEE Trans. Biomed. Eng. **59**(9), 2677–2683 (2012)
10. X. Glorot, Y. Bengio, Understanding the difficulty of training deep feedforward neural networks, in *Proceedings of the 13th International Conference on Artificial Intelligence and Statistics* (2010), pp. 249–256
11. M.N. Gurcan, L. Boucheron, A. Can, A. Madabhushi, N. Rajpoot, B. Yener, Histopathological image analysis: a review. IEEE Rev. Biomed. Eng. **2**, 147 (2009)
12. N.A. Hamilton, R.S. Pantelic, K. Hanson, R.D. Teasdale, Fast automated cell phenotype image classification. BMC Bioinform. **8**(1), 110 (2007)
13. V. Iglovikov, S. Mushinskiy, V. Osin, Satellite imagery feature detection using deep convolutional neural network: a Kaggle competition (2017), arXiv:1706.06169
14. V. Iglovikov, A. Shvets, TernausNet: U-Net with VGG11 encoder pre-trained on ImageNet for image segmentation (2018), arXiv:1801.05746
15. M. Kowal, P. Filipczuk, A. Obuchowicz, J. Korbicz, R. Monczak, Computer-aided diagnosis of breast cancer based on fine needle biopsy microscopic images. Comput. Biol. Med. **43**(10), 1563–1572 (2013)
16. A. Krizhevsky, I. Sutskever, G.E. Hinton, ImageNet classification with deep convolutional neural networks, in *Advances in Neural Information Processing Systems* (2012), pp. 1097–1105
17. Y. Li, J. Wu, Q. Wu, Classification of breast cancer histology images using multi-size and discriminative patches based on deep learning. IEEE Access **7**, 21400–21408 (2019)
18. G. Litjens, C.I. Sánchez, N. Timofeeva, M. Hermsen, I. Nagtegaal, I. Kovacs, C. Hulsbergen-Van De Kaa, P. Bult, B. Van Ginneken, J. Van Der Laak, Deep learning as a tool for increased accuracy and efficiency of histopathological diagnosis. Sci. Rep. **6**, 26286 (2016)
19. A. Rakhlin, A. Shvets, V. Iglovikov, A.A. Kalinin, Deep convolutional neural networks for breast cancer histology image analysis, in *International Conference Image Analysis and Recognition* (Springer, 2018), pp. 737–744
20. S. Robertson, H. Azizpour, K. Smith, J. Hartman, Digital image analysis in breast pathology–from image processing techniques to artificial intelligence. Transl. Res. **194**, 19–35 (2018)
21. J. Schmidhuber, Deep learning in neural networks: an overview. Neural Netw. **61**, 85–117 (2015)
22. Y. Shin, I. Balasingham, Comparison of hand-craft feature based SVM and CNN based deep learning framework for automatic polyp classification, in *2017 39th Annual International Conference of the IEEE Engineering in Medicine and Biology Society (EMBC)* (IEEE, 2017), pp. 3277–3280
23. K. Simonyan, A. Zisserman, Very deep convolutional networks for large-scale image recognition (2014), arXiv:1409.1556
24. F.A. Spanhol, L.S. Oliveira, C. Petitjean, L. Heutte, A dataset for breast cancer histopathological image classification. IEEE Trans. Biomed. Eng. **63**(7), 1455–1462 (2015)
25. F.A. Spanhol, L.S. Oliveira, C. Petitjean, L. Heutte, Breast cancer histopathological image classification using convolutional neural networks, in *2016 International Joint Conference on Neural Networks (IJCNN)* (IEEE, 2016), pp. 2560–2567
26. C. Szegedy, W. Liu, Y. Jia, P. Sermanet, S. Reed, D. Anguelov, D. Erhan, V. Vanhoucke, A. Rabinovich, Going deeper with convolutions, in *Proceedings of the IEEE Conference on Computer Vision and Pattern Recognition* (2015), pp. 1–9

27. D. Wang, A. Khosla, R. Gargeya, H. Irshad, A.H. Beck, Deep learning for identifying metastatic breast cancer (2016), arXiv:1606.05718
28. S.C. Wong, A. Gatt, V. Stamatescu, M.D. McDonnell, Understanding data augmentation for classification: when to warp? in *2016 International Conference on Digital Image Computing: Techniques and Applications (DICTA)* (IEEE, 2016), pp. 1–6
29. Y. Yu, H. Lin, Q. Yu, J. Meng, Z. Zhao, Y. Li, L. Zuo, Modality classification for medical images using multiple deep convolutional neural networks. J. Comput. Inf. Syst. **11**(15), 5403–5413 (2015)

Chatterbot: Technologies, Tools and Applications

Gajendra Kumar Ahirwar

Abstract Chatbot is used to chat with live user and assist human to seek information and perform skilled task instantly. Later then many chatbot came, some of the voice-based chatbots (as like assistant) are SIRI, CORTANA, BIXBY, ALEXA, NATASHA, IBM WATSON and many more. There was no game changing in the chatbot technologies, it is obvious that they evolved from the very simple pattern matching systems towards complicated patterns combined with a set of concept and categories in a subject area or domain in which the knowledge is enabling computer reasoning. This application working is very simple because it is already known in advance. The techniques which will be improving the chatbots are artificial intelligence, machine learning with natural language processing. This chapter proposal is representative of the significant technologies, tools and methods in Chatbots in the last decade.

1 Introduction

Chatbots or Chatterbot are smartly written computer program that are capable to carry communication between users. They are automated computer programs that perform communication. The main aim of making chatbot is that they give us the feed of human communication. They just simply understand the language that we use in out chats, analyse it and logically give the conclusion [1].

In other words we can say that a chatbot is an artificial intelligent program which is used to carry near and natural conversation. The input to this program is natural language text, and the application should give an answer that is the best intelligent response to the input sentence. According to the database with which they are connected give the logical smart and intelligent answers. In 1966, ELIZA is the first chatbot that was introduced by Joseph Weizenbaum [2].

G. K. Ahirwar (✉)
Department of Information Technology, Rajiv Gandhi Proudyogiki Vishwavidhyalaya Airport Road Gandhi Nagar, Bhopal, Madhya Pradesh, India
e-mail: gajendrakumarahirwar@gmail.com

A. Nanda and N. Chaurasia (eds.), *High Performance Vision Intelligence*,
Studies in Computational Intelligence 913,
https://doi.org/10.1007/978-981-15-6844-2_14

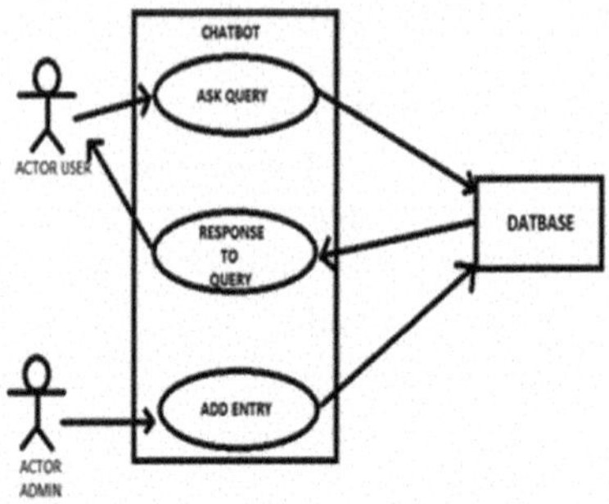

Fig. 1 Chatbot working diagram

Chatbot design techniques in the last decade from the survey, it can be said that the development and improvement of Chatbot design is not growing as an organization rate due to the variety of methods and approaches used to design a Chatbot. There is some truth in it, however, it is conspicuous that the recent developments especially the chatbots are moving out of the scripted era. It is obvious that there is a trend towards the study and meaning which can lead to a conclusion that future chatbots will evolve from pattern matching. General-purpose Chatbots need improvements by designing a more comprehensive knowledge base. Turing test is also used for improving the skill of chatbot for playing the games [3] (Fig. 1).

2 Applications of Chatterbot

Following covers the studied applications of the Chatterbot.

2.1 Customer Support System

Chatbots provide support to their customers who need help in completing their actions [4]. It also helps in finding the products having discounts. Sometimes the customer is looking for the product with the discounts so it help to do so.

2.2 Suggest Product

It helps the customer in finding the specific product. Many a times a customer wants to buy a product but does not buy it may be it is not getting the product to his satisfaction so here chatbot will help one to find specific.

2.3 E-Commerce

Chatbots add new layer of interactivity to e commerce. The major use cases for e commerce are as follows:

(a) set price alerts
(b) order physical goods
(c) buy gifts
(d) reserve services

2.4 Travelling and Vacations

Chatbots help in booking travel. It is also helpful in solving travel-related problems. As we know that vacation planning is one of the most frustrating task because of choosing best vacation place, country and some other information. Now with the help of chatbot the user select the best vacation or travelling information.

2.5 Queries and Complaints

Chatbot helps the dissatisfied customers by surveying the queries and complaints [5]. Chatbot applications pick up the query and find out the best answer which satisfied the given query or problem. It is worked based on natural language processing.

3 Applications of Chatterbot

The following section discusses about the studied related work on the subjected title of the chapter.

3.1 Eliza

Eliza is the first text-based chatbot that was introduced in 1964 by Joseph Weizenbaum.Eliza use keywords to determine the user requirements and to understand, it also set the transformation rules for output [1].

3.2 Parry

In 1972, psychiatrist Kenneth Colby designed the parry chatbot at Stanford university. In parry they included many advanced features than Eliza. It was described as a "Eliza with attitude" [6].

3.3 Jabberwacky

The main aim of Jabberwacky chatbot is "simulate natural human chat in an interesting, entertaining and humorous manner" [7, 8].

3.4 Alice

ALICE (Artificial Linguistic Internet Computer Entity) created by Richard Wallace in 1995ALICE is stand for Artificial Linguistic I [9].

3.5 Watson

IBM developed the 'WATSON' chatbot in 2006. Watson uses IBM's DeepQA software and the Apache UIMA (Unstructured Information Management Architecture) framework. The main aim IBM Watson is defeating the best human players in February 2011 [7, 9].

3.6 Mitsuku

Mitsuku is a female chatbot of 18 years old. It is based on artificial intelligence markup language (AIML).

3.7 Siri

Siri is automatic speech recognition(ASR) to convert human speech into text. It uses natural language processing (NLP) to transcribed text into "parsed text". SIRI works as a personal assistant [10].

3.8 Alexa

Alexa [11] is an intelligent personal assistant developed by Amazon. It works on voice integration, music play by making of to do list, setting alarm, streaming podcasts and playing audio.

3.9 Tay

Tay is developed bu Microsoft in 2016. It is an artificial intelligence chatterbot that calls controversy on twitter by reducing in flammatory tweets.

4 Analysis of Different Chatbots

The Table 1 shows the feature analysis of different chatbots.

5 Technology Used in Chatterbot

There are different technologies used in chatbot like Artificial intelligence, Natural language processing, Machine learning, etc.

5.1 Artificial Intelligence

Artificial Intelligence is the investigation of how to cause PCs to accomplish the thing which right now individuals improve.

Artificial intelligence is a field of computer science which is developed in 1956 by American computer scientist 'John McCarthy'. Artificial Intelligence is a unique field of Information Technology which is growing rapidly. It is a technology which helps to make a machine capable to perform a task as like as a human being. It provides the tools and techniques to develop an automated system which work like human beings. AI helps to understand the human brain and try to solve the queries that come in human mind. Artificial intelligence basically has the following goals.

(a) Acting Humanly which is based on turing test approach.
(b) Acting Rationally which is based on rational agent approach.
(c) Thinking humanly which is based on cognitive modelling approach.
(d) Thinking Rationally which is based on law of thought approaches.

Table 1 Analysis of different chatbots with their methodology and key points

S. No.	ChatterBot	Year	Developed by	Methodology	Key point
1	ELIZA	1966	Joseph Weizenbaum	Pattern matching and substitution methodology	Recognizing keywords
2	PARRY	1972	Kenneth Colby	Natural language processing	Assumption, attribution, emotional response
3	JABBER WACKY	1988	Rollo Carpenter	Contextual pattern matching	Individual use, human conversation
4	Dr. SBAITSO	1992	Creative labs for MS-DOS	Artificial intelligence	Text to speech
5	ALICEBOT	1995	Richard Wallace	Heuristic pattern matching, AIML	Open source, chatting, conversion
6	SMART CHILD	2001	Active buddy	Natural language	Fun conversation, MSN messenger
7	WATSON	2006	IBM	AI	Defeating best human player
8	MITSUKU	2013	Steve Worswick	AIML	18 Year female Chatbot
9	SIRI	2011	Apple Inc.	Natural language UI	Voice based chatbot, navigating areas, scheduling events and reminders
10	GOOGLE NOW	2012	Google Inc.	Natural language processing, artificial intelligence	Voiced portable assistant, answer questions
11	CORTONA	2014	Microsoft Inc.	Voice Recognition	Reminder based on time chit-chat
12	ALEXA	2014	Amazon	Alexa skills kit (ASK)	Virtual assistant, search web, create to-do
13	TAY	2016	Microsoft Corporation	Artificial intelligence	Interact with twitter human users

AI use the different intelligent behaviours like perception, reasoning, learning, communicating, etc. The main goal of artificial intelligence is to develop the machines much intelligent as well as humans.

There are various branches of artificial intelligence.

1. Planning
2. Searching
3. Learning from Experience
4. Reasoning
5. Heuristics
6. Representation
7. Pattern Recognition
8. Game Playing
9. Computer Visions
10. Expert System

The goals of AI researchers are to develop the successful methods for dealing uncertain and incomplete information. To solve any problem through artificial intelligence, the AI programs use the reasoning, judgements and sensor skills like humans so machines provide the fast and required efficient output in minimum time.

Intelligent programmers firstly set the goals for the problems than achieve the solution or goals to the particular problem with different planning and learning techniques. In planning predict the different states of the problem and changes them according to the needs while the learning is a activity to gain the knowledge. Basically learning is divided into two main categories in artificial intelligent. First one is supervised learning which works with teacher and second is unsupervised learning which works without the need a teacher.

Artificial intelligent-based program uses the heuristic search technique. The solution steps are not explicit and the knowledge is imprecise for AI programs while the conventional program used the algorithmic search technique.

The artificial intelligence has some limitations as low perception, low self decision and low thinking capacity.

5.2 Machine Learning

A Machine Learning is a subject that provides an algorithm for understanding the human mind. Machine Learning helps to predict the human mind by their previous activities. Machine Learning is a modal is used to know the mind of people. It predicts the possibilities and provides the best option according to their knowledge. Machine Learning helps to learn and experience and to increase the quality of product. It aims to focus on accuracy rather than the chance of success. Machine goes for solution is developed by Richard Wallace in 1995. Artificial Linguistic Internet Computer Entity (ALICE) is based on Natural Language Processing chatbot. It matches heuristical patterns matching rules to humans input. Whether it is optimal or not. Machine Learning tries to minimize the complex problem.

Machine Learning is a subset of man-made reasoning which centers primarily around machine learning from their experience and making forecasts dependent on its experience.

Machine learning has the various techniques for making the intelligence devices. Some of them are as follows:

- Genetic Algorithms
- Artificial Neural Network
- Bayesian learning
- Computational learning theory
- Instance-based learning
- Analytical learning
- Reinforcement learning

There are basically three types of machine learning.

1. Supervised learning
2. Unsupervised learning
3. Reinforcement learning

The main issue in machine learning is that which algorithm is best for which types of problems and how to determine that how much training set is sufficient.

5.3 *Natural Language Processing*

'Natural Language Processing is a ability of a computer to understand what a human is saying to it.'

Natural language processing (NLP) is a branch of artificial intelligence in which natural language is used to communication with computer system. The natural language is generally a language that is used by humans to communicate with each other [12]. Natural language processing automatic processes the human language thus it performed the human–machine communication. The sentences (strings set) are used as an input in natural language processing and produce the structured representation of those strings. The first natural language was manipulated by the computer in 1950, in which automatic translation was done between Russian and English. There are basically two main problems in natural language processing.

- Ambiguity Level
- Complexity of semantic information
- Functional structure
- Phrases repetition

The natural language processing or NLP has the following phases:

1. **Morphological Processing**
   In this phase, the sentences are divided into strings and strings are broken in sub-words, punctuations, tokens for language processing.
2. **Analysis of Syntax**
   In this phase, the sentences are checked. It consists of the syntax or grammar rules for sentence checking.
3. **Semantic Analysis**
   This stage analyses the grammar of the sentence. In other words, we can say that at this stage the meaning of the words is checked.
4. **Discourse Integration**
   Sentences are interpreted in this phase.
5. **Analysis of Pragmatic**
   In this phase, it is checked that how to use a sentence in different situations.

NLTK [13], is used to deal with natural language processing. It is a open source tool. It used as a free plugin in python programming language.

5.4 Knowledge Representation

'Knowledge representation is a field of artificial intelligence, in which the main goal of knowledge representation is how to represent the knowledge to solve the problems.' In knowledge representation there are two basic entities exist.

1. **Facts**
   Facts show a particular situation. The statements are verified by it.
2. **Representation of facts**
   Facts are represented in such a way that the problems can be easily understood and manipulated.

In knowledge representation the main focus on facts, representation and their mapping. There are two types of representation mapping.

- Forward representation mapping
- Backward representation mapping

Knowledge representation has four main properties or approaches in a particular domain which one representation adequacy, inferential adequacy, inferential efficiency and acquisition efficiency. The fundamental goal of knowledge representation is to achieve a conclusion from the knowledge base. Various issues generated when knowledge representation techniques are used. Some of them are as follows.

- Attributes
- Relationship among attributes
- Finding right structure
- Sets of objects
- Choosing granularity

6 Tools for Chatbots

Chatbots has various tools for its developing and working. Here present the main three tools.

6.1 Chatful

Chatful is a platform used for building the different types of chatbots or chatterbot. It helps us in creating Artificial intelligent-based chatbot on facebook messenger without IT Skills. It provide Artificial intelligence (AI) automation which help the bot to understand the chat which is send by the users [14].

6.2 Botsify

Botsify is used to create the chatbot for different Websites. It allows us to collect users data. All the data can be stored in dashboard, send via E-mail or exported to CSV file. It also helps your bot in learning new keywords, terms and pharses

6.3 Smooch

Smooch is a multichannel platform tool for chatbot. It connects the different messaging applications in a single interface or API. This tool provides the plateform for various chatting applications to integrated in a single place so the users from different messaging channels connect together into a single api.

6.4 Pandorabot

Pandorabot [15] is a free IDE for developing the chatbots in easy way. It has the different features and many tools for basic building blocks of bot.

7 Conclusions

With the rapid development of World Wide Web, it is becoming more difficult for any user to access the required data more quickly. To get data quickly we need to use speech which is one of the most powerful forms of communication with human beings. The crux of this paper is to prepare comprehensive comparison of chatbot systems write from the first chatbot ELIZA to one of the latest chatbot like ALEXA. Artificial Intelligence chatbot or conversational agents can be used to automate the interaction between user and the device. The chatbot is the virtual communication between the human and machine; Nine studies that made perceptible contributions in Chatbot design in the last ten years which are selected and then, reviewed. The different techniques, tools and applications of chatbots are reviewed.

Acknowledgements I would like to thank my supervisor who has supported and provided guidance to me whenever required.

References

1. Z. Darius, S. Hundertmark, chatbots - an interactive technology for personalized communication, transactions and services. IADIS Int. J. WWW/Intern. **15**(1), 96–109 (2018)
2. J. Weizenbaum, ELIZA - a computer program for the study of natural language communication between man and machine. Commun. ACM **9**(1) (1995)
3. P. Hingston, A turing test for computer game bots. IEEE Trans. Comput. Intell. AI Games **1**(3), 169–186 (2009)
4. E.H. Almansor, F.K. Hussain, *Survey on Intelligent Chatbots: State-of-the-Art and Future Research Directions*, ed. by L. Barolli et al., CISIS 2019, AISC 993 (2019), pp. 534–543, https://doi.org/10.1007/978-3-030-22354-0_47
5. A. Mishra, S. Sapre, S. Shinde, S. Nahar, S.N Shelke, Intelligent chatbot for guided navigation of repository contend. Int. J. Adv. Res. Comput. Commun. Eng. **8**(5) (2019)
6. A.I. Alice, Foundation. Free A.L.I.C.E. AIML Set (2015) March 21, http://code.google.com/p/aiml-en-us-foundation-alice/
7. J. Joab, *IBM Watson Vanquishes Human Jeopardy Foes* (PC World, IDG News, 2011)
8. Jabberwacky IEEE 802.3 12.4.3.2.3 Jabber function
9. P. Hingston, A turing test for computer game bots. IEEE Trans. Comput. Intell. AI Games **1**(3), 169–186 (2009)
10. B. Abu Shawar, E. Atwell, Chatbots: are they really useful? LDV-Forum 2007 **31** (2007)
11. Amazon Developer, Alexa (2017), https://developer.amazon.com/alexa
12. S.A. Abdul-Kader, J. Woods, Survey on chatbot design techniques in speech conversation systems. (IJACSA) Int. J. Adv. Comput. Sci. Appl. **6**(7) (2015)
13. S.A. Abdul-Kader, J. Woods, Survey on chatbot design tech-niques in speech conversation systems. (IJACSA) Int. J. Adv. Comput. Sci. Appl. **6**(7), 72–80 (2015)
14. K. Nimavat, T. Champaneria, Chatbots: an overview types, architecture, tools and future possibilities. IJSRD - Int. J. Sci. Res. Develop. **5**(07) (2017)
15. N. Jain, A. Jain, A survey on popularity of chat-bots **5**, 277–280 (2017)

Route Planning Using Nature-Inspired Algorithms

Priyansh Saxena, Raahat Gupta, and Akshat Maheshwari

Abstract There are many different heuristic algorithms for solving combinatorial optimization problems that are commonly described as Nature-Inspired Algorithms (NIAs). Generally, they are inspired by some natural phenomenon, and due to their inherent converging and stochastic nature, they are known to give optimal results when compared to classical approaches. There are a large number of applications of NIAs, perhaps the most popular being route planning problems in robotics—problems that require a sequence of translation and rotation steps from start to the goal in an optimised manner while avoiding obstacles in the environment. In this chapter, we will first give an overview of Nature-Inspired Algorithms followed by their classification and common examples. We will then discuss how the NIAs have applied to solve the route planning problem.

1 Introduction

Man has always looked toward nature as an origin of insight to answer the most complex of questions that surround us. Many complex physical phenomena that first seem mystifying or baffling at least, can be found already being produced and controlled in nature. A careful study of these can lead to the development of algorithms and theories can be used to solve complex problems in our macro world. The techniques that are inspired either directly or indirectly from processes observed in nature are called **Nature-Inspired Algorithms**, and their applications in effective route planning will be the main focus of this chapter.

P. Saxena (✉) · R. Gupta · A. Maheshwari
ABV-Indian Institute of Information Technology and Management, Gwalior, India
e-mail: saxenapriyanshasd@gmail.com

R. Gupta
e-mail: raahat.gupta.1998@gmail.com

A. Maheshwari
e-mail: aks3d76@gmail.com

A. Nanda and N. Chaurasia (eds.), *High Performance Vision Intelligence*,
Studies in Computational Intelligence 913,
https://doi.org/10.1007/978-981-15-6844-2_15

Every natural organism living today has gone through centuries and millennia of evolution. This has lead to the development of those body organs which are essential in the survival of the species in the long run, for example, turtles developed shells to protect against other creatures, birds evolved with wings to fly, and apes and early humans started using their front limbs to hold things. This several hundreds of millions of years of evolution are the primary reason we look for inspiration in the first place—for instance, we can solve complex transportation problems in toady's world by taking a closer look at how ants and other species look for their food in the wild.

This chapter takes a more in-depth look at various Nature-Inspired Algorithms (NIAs) and how we can apply them in optimisation problems, specifically route planning. We will also discuss the walk-through of a basic NIA, along with giving examples of several NIAs used in today's world. This chapter concludes by giving some insights into the motivation of taking inspirations from nature.

2 The Problem of Route Planning

Humans have reached a great deal of automation in the twenty-first century. Examples of programmed robots range from the ones operating forklifts in Amazon warehouses, to aerial drones supplementing the delivery of medical supplies in Rwanda and Kenya, and rovers on the surface of Mars, Moon and other celestial bodies. All of these robots work in complex terrains, and they have to be programmed to automatically detect obstacles in their route, and take measures to avoid them.

The gist of the **route planning** problem is as follows—the agent/robot has to move from a starting location to its goal location while avoiding any/all obstacles in its route. Since obstacles in real world can be, and often are, random and arbitrary, their positions cannot be accurately defined beforehand. Thus the robot has to *search* for obstacles and develop alternate routes to avoid them on-the-fly. This is what makes the problem of route planning so tricky (it is actually NP-hard), and its immense applications in today's world mean that route planning has become a hot research topic in recent years.

A formal definition of route planning is given in the following section.

2.1 Overview of Route Planning

Route planning is considered as the process of identifying a sequence of transnational and rotational steps that the robot must take to reach destination location from a given source location in the shortest possible time and by avoiding collisions within the environment which contains static as well as moving obstacles. This problem becomes exceedingly challenging when the environment is made of dynamic obstacles because it requires the route to be re-planned in real time when a new obstacle in discovered. Additional complexities include that the route should be simple, smooth and should be realistic to the degree that it can be followed by an unmanned aerial vehicle (UAV) in a practical scenario. Researchers generally consider two types of robots for route planning: *land robots* are generally restricted to two-dimensional environments. On the other hand, *air and water robots* have more degrees of freedom and can change their position in any direction in space. Due to their complexity and immense practical applications, aerial robots have been studied the most.

Route-planning approaches can be classified on the basis of two approaches: (i) **global route planning** also called *offline approach*, and (ii) **local route planning** or *online approach* [1]. In global route planning, we create a high-level route based on a map whose current and past perceptive are known to the algorithm. It generally produces a low-dimensional and optimised route; however, it fails to react to unknown obstacles. Online route planning algorithms do not input any previous information based on the environment. The route it gives, which is high-dimensional and low-level, is only over a fragment of the total environment. This fragment can be re-routed when a new obstacle arrives without altering the entire route. Hence, local planners are more suited for highly changing environments.

2.2 Methods to Solve Route Planning Problems

Several methods have been proposed to provide solutions for route planning problems. These are classified into: *classical methods* and *heuristic methods*.

2.2.1 Classical Methods

Many classical methods for route planning have been proposed over the decades. Here, we will discuss four such methods: Cell Decomposition method (CD), Potential Field Method (PFM), Sub-goal method (SG) and Sampling-based methods.

Cell Decomposition Method This method is based on the process of dividing the available free space in a robot's configuration into smaller regions, also known as *cells*. This process gives a simple route collision-free route [2].

Potential Field Method This method takes inspirations from the concept of potential field in electrostatics. The agent is analogous to a charged particle, and the goal is assigned an attractive charge. Thus, the agent will be attracted towards the goal. The obstacles, on the other hand, are assigned a negative force with respect to the agent. Thus, the agent will be repelled from the obstacles and move towards the goal [3]. The general formula for finding repelling potential is altered to give a reduced number of oscillations and avoiding improving practical concerns. An advantage of this method is that it can be successfully extended to multiple cruise UAVs as well.

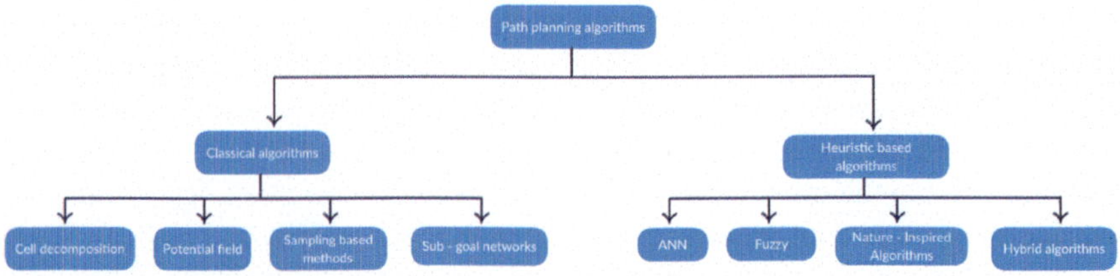

Fig. 1 The classification of robot route planning algorithms

Sub-goal Method This method makes use of different attainable configurations from start to the goal, while bypassing all obstacles. This method is immensely popular for applications in robot navigation, as seen in [4].

Sampling-Based Motion Planning These schemes are especially useful in complex real-life planning scenarios. The most influential sampling-based methods consist of *Rapidly-exploring Random Trees (RRT)* and *Probabilistic Road-Map (PRM)* [5]. The similarity between these two methods is the concept of randomly sampled connecting points, however, they are different in the aspect that the graph which connects the points is created.

Constant efforts are being made for applying classical methods to solve practical and real-world problems. Some of the algorithms that have shown promise include *Configuration Space Collision Maps* and *Voronoi Diagrams*. The working of these algorithms is left to the reader as a further study [6–8]. Disadvantages of classical approaches include their non-ability to produce optimal routes and tendency to get stuck in local minima. The existence of multiple obstacles provides a hindrance to well functioning of some of the environments. Hence, heuristic approaches are more prevalent today (Fig. 1).

2.2.2 Heuristic Methods

Heuristic methods are immensely popular in the field of route-planning research toady. Some of the standard methods are discussed below.

Neural Networks The various algorithms that come under the umbrella of neural networks have seen an explosion in popularity in the past few years. These methods have also been applied to solve route planning problems [9]. Neural networks are known for finding out the relations between inputs and outputs. Their working, especially in robot navigation, is categorised as: (i) interpreting input data, (ii) enforcing techniques of obstacle avoidance, (iii) implementing route planning. A common neural network paradigm used for training **Recurrent Neural Networks (RNN)** is *Reservoir Computing*. The model is made up of two randomly generated RNNs—one of which models the localisation ability and the other "learns" the navigation skill. A collision-free route can also be constructed using a combination of two neural network concepts. These are *principal component analysis (PCA)* and a *multi-layer perceptron (MLP)*.

Fuzzy Logic The robot takes decisions with reference to a set of IF-THEN rules. It starts by diving the task into simpler problems [10]. The cost function is calculated and goals such as orientation of target, avoidance of obstacles and rotational movement are included in the above-mentioned cost function in order to determine the optimum steering angle θ. The mobile robot varies the weights of the cost function and navigates through the environment.
Hybrid Algorithms Researchers have integrated neural networks with their fuzzy counterparts in an effort to extract the positives of both the algorithms. This hybrid is also known to show positive results.
Nature-Inspired Algorithms NIAs are the most recent addition to this vast list of algorithms. These are generally more complex than the above-discussed algorithms and required more time and computation power. However, NIAs are shown to give more optimised and accurate answers than the above algorithms. As a result, NIAs will constitute a major portion of the remainder of this chapter. Most NIAs include the concept of **Swarm Intelligence (SI)**, which is the property of a system where the composite demeanour of unsophisticated agents interacting locally with their environment cause orderly functional global patterns to emerge [11]. Examples of NIAs may include Particle Swarm Optimisation (PSO), Salp Swarm Algorithm (SSA), Artificial Bee Colony Optimisation(ABC) and Ant Colony Optimisation(ACO) among others. These algorithms, along with the concept of Swarm Intelligence, will be discussed in further depths in the upcoming sections.

3 Nature-Inspired Algorithms

The scientific field of bionics helps us link biological processes, functions and organisational principles to modern technologies. There is a diverse range of algorithms, mathematical and meta-heuristic, developed specifically for the process of assigning expertise from the biological lifeforms to human life through technology. As a result, different kinds of optimisation algorithms have been developed which apply the concepts found in nature to practical and research life [12]. The techniques - Nature-Inspired Algorithms—will be the focus of this section.

3.1 Objectives of NIA

NIAs are designed with an objective of optimisation—i.e., of a problem of which multiple solutions exist, and the interest is to determine the solution with minimum cost. Usually, the cost here is the lowest time taken, but other factors such as lowest distance, or a combination of the two can be considered.

Here, we describe two key factors terms useful to understand the constraints of most optimisation problems: **exploration** and **exploitation**. Exploration consists of finding the global optima by searching the entire search space; as the name suggests, we are *exploring* our surroundings in hopes of finding resources that are better than what we currently have. Exploitation, on the other hand, consists of finding the local optima in the explored solution space. We are *exploiting* what all resources we have to get the best possible output. It is important to understand that these two processes are often applied simultaneously—intense exploration does not give optimal solution while deep exploitation traps an algorithm in local optima. Hence, a balance between these two processes is necessary [13].

Nature-Inspired Algorithms are broadly classified into two categories: *Genetic Algorithms (GA)* and *Swarm Intelligence (SI)*. These two categories are described in the next sections.

3.2 Genetic Algorithms

Genetic Algorithm (GA) was developed as a tool to understand the natural processes behind millions of years of evolution. It involves a study of processes like natural selection, crossover, and mutation, hence, they are also known as *evolutionary algorithms*. Later these concepts are applied to optimisation problems and have their applications in route planning.

The major inspiration behind GA has been the theory "Survival of the Fittest" given by Charles Darwin [14]. These algorithms are able to apply natural selection, recombination and mutation so as to closely resemble the processes by living organisms, while at the same time giving impressive results in modern research.

3.3 Swarm Intelligence

Swarm Intelligence (SI) mainly composes of three principles: *evaluation*, *comparing* and *imitation*. The term evaluation denotes the capability to analyse the positive and negative effects of property in nature. Comparing comes naturally to living beings, wherein they compare themselves with other beings. The main purpose of these comparisons is to bring a sense of motivation to learn and/or modify. Imitation is defined as an effective form of learning.

Properties of a Swarm Intelligence System Following are the properties generally seen in a Swarm Intelligence (SI) system:

1. A swarm is generally composed of many individual entities. Hence, the effects of outliers are diminished.
2. The individuals that make up the swarm are generally homogeneous—they are either identical or belong to only a few classifications.
3. Individuals interact among each other and share information about themselves or what they have learnt about the environment; this sharing is either direct or indirect.

4. These interactions between the individual "particles" constitute the overall behaviour of the system—the group behaviour self-organises.

Principles Involved in Swarm Intelligence Swarm Intelligence can be described by taking into consideration the following principles [15]:

1. **Proximity Principle**: the individual should be placed such that it can perform simple space and time computations.
2. **Quality Principle**: the individual must respond to and satisfy various quality factors specified in the environment.
3. **Diverse Response Principle**: the individual should not execute its activity along extremely limiting channels.
4. **Stability Principle**: the individual should never change its style of action even though the environment may change.
5. **Adaptability Principle**: the individual must have the ability to change its behaviour mode should the computational price justifies it doing so.

4 Walk-Through of an NIA

In this section, we will discuss in detail the formulations of a general Swarm Intelligence NIA, and how it is applied in practice to tackle optimisation problems such as route planning.

There are many NIAs which are used for route planning in practice. One of the recent and most effective ones is called *Salp Swarm Algorithm (SSA)*, proposed by Mirjalili et al. in 2016 [16]. It is an efficient algorithm that depicts swarm intelligence by mimicking the behaviour of groups of swarms, which are a sea creature similar to jellyfish. Although it is relatively easy to understand, SSA is an effective algorithm for solving real-world problems. The problems tend to have unknown and challenging search spaces.

4.1 The Inspiration Behind Salp Swarm Algorithm

Salps are a sea creature. They are members of the *Salpidae* [16] family. Salps are known to have a barrel-shaped transparent body similar to jellyfish. Their movement can also be compared to that of jellyfish—they both pump water through their body and move forward as a result of the propulsion.

Salps are known to exhibit *swarming behaviour*, which will be of particular interest to us in this chapter. A group of salps, which is formed during swarming, is known simply as a **salp chain**. The primary motivation behind this swarming behaviour is unclear, but researchers speculate the motive may be better locomotion of an otherwise vulnerable creature in a hostile environment [17].

4.2 Mathematical Model of SSA

To mathematically model, a natural phenomenon means to devise mathematical equations based on the observations and use these equations in software simulation programs. Before applying SSA, we first divide the salp "population" in two groups: *leaders* and *followers*. The salp at the frontal end of the chain is termed as leader. On the other hand all the other salps that follow the leader are called followers. The job of the leader is to guide the salp chain and the others directly (or indirectly) follow their leader.

We use an m-dimensional search space to characterise the locale of salps, where m is the number of variables given. We also assume that a two-dimensional matrix, x, is used to store the position of all salps. If the target of the entire chain is defined by a food source, $Food$, then the steps to update the leader's position is given in Eq. 4.1

$$x_j^1 = \begin{cases} Food_j + r_1((upper_j - lower_j)r_2 + lower_j), & \text{if } r_3 \geqslant 0, \\ Food_j - r_1((upper_j - lower_j)r_2 + lower_j), & \text{if } r_3 < 0, \end{cases} \tag{4.1}$$

where the position of the leader in the jth dimension is given as x_j^1, the food source is given as $Food_j$, and $upper_j$ and $lower_j$ are the respective upper and lower bounds of the jth dimension. r_1, r_2 and r_3 are three random numbers into the equation.

It can be inferred from Eq. 4.1 that only the leader updates its position regarding the food source. The most important parameter in SSA is actually r_1 as it can be adjusted to balance *exploration* and *exploitation*. r_1 can be obtained from Eq.'4.2

$$r_1 = 2e^{-(\frac{4n}{N})^2}, \tag{4.2}$$

where N is the maximum number of iterations and n is the current iteration.

The other random numbers in Eq. 4.1, r_2 and r_3, are uniformly generated in an interval of [0,1]. They define whether the upcoming position in jth dimension would be towards positive direction, or negative direction. They also indicate the step size.

The position of the followers is updated by utilising Newton's laws of motion, Eq. 4.3:

$$x_j^1 = \frac{1}{2}at^2 + v_0 t, \tag{4.3}$$

where x_j^i is the position of ith follower salp in jth dimension, for all $i \geqslant 2$. Here, t is time and v_0 defines the initial speed. The acceleration, $a = \frac{v_{final}}{v_0}$, and velocity, $v = \frac{x - x_0}{t}$.

This Eq. 4.3 is modified relative to the SSA, and can be written as

$$x_j^i = \frac{1}{2}(x_j^1 + x_j^{i-1}), \tag{4.4}$$

where x_j^i shows the position of ith follower salp in jth dimension, for all $i \geqslant 2$.

The positions of all the salps can be simulated now by using Eqs. 4.1 and 4.4.

4.3 *Working of SSA*

To delve into the inner workings of Salp Swarm Algorithm, it is important to realise that the leader salp alone 'leads' the chain and the other salps simply 'follow'. The salps chase the food source. So, if we replace the food source with global optima, then the salp chain automatically moves in that direction. The pseudocode of SSA algorithm can be stated as

Algorithm 2: SSA Algorithm

Initialise the salp population x_i (i = 1, 2, ..., m) considering *upper* and *lower*
while (end condition is not satisfied)
Calculate the fitness of each search agent (salp)
$Food$ = the best search agent
Update r_1 by Eq. 4.2
    for each salp (x_i)
        if (i == 1)
            Update the position of the leading salp by Eq. 4.1
        else
            Update the position of the follower salp by Eq. 4.4
        end
    end
    Amend the salps based on the upper and lower bounds of variables
end
return $Food$

We start the algorithm with multiple salps in random positions. We then calculate the *fitness* of each salp. The best-fitness salp is assigned to the variable $Food$, and it becomes the food source which is then chased by the salp chain. Now, we update the parameter r_1 using Eq. 4.2. Subsequently, the position of leader salp is updated using Eq. 4.1. Similarly, the position of the follower salp is updated using Eq. 4.4, as described in Sect. 4.2. We perform boundary check at every step, wherein we check if any salp goes outside the boundaries of the environment. If it does, it is brought back inside the boundary. We execute all the other steps iteratively, until a suitable satisfaction condition is reached.

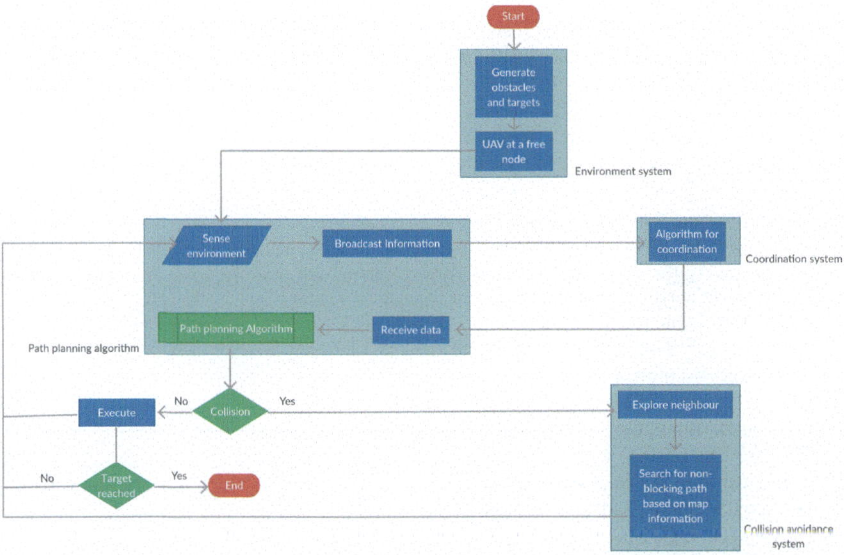

Fig. 2 Methodology of applying SSA to route planning

One thing to note here is that salp chain can determine a better solution to the problem through exploration and exploitation. Hence, the food source is updated during this process. Thus, the salp chain should be able to chase a moving food source, making it useful in a dynamic target environment.

4.4 Route Planning Using SSA

Figure 2 depicts the methodology undertaken when applying SSA to route planning. In the route planning phase, the robot will sense the environment and broadcast this information to the semi-centralised server, which is employed for achieving coordination task among the robots. The received data from the server is used for route planning and moving towards the target. If there is no collision detected, the robot is moved to the required position and checked for target reachability. If the collision is detected collision avoidance algorithm is run, which includes a different strategy for dealing with the static and dynamic obstacles. After collision avoidance, the route planning system is kicked again, and it continues.

The results obtained by applying SSA to 3D static environments are depicted in Fig. 3. Here, the coloured blocks represent the obstacles of varying shapes and sizes, and the agent moves from left to right, and the blue line traces its route.

SSA for dynamic environment is depicted in Fig. 4. Here, the grey blocks are static obstacles and the black dots contain dynamic obstacles. The red dots represent the targets and the route is traced by the blue line.

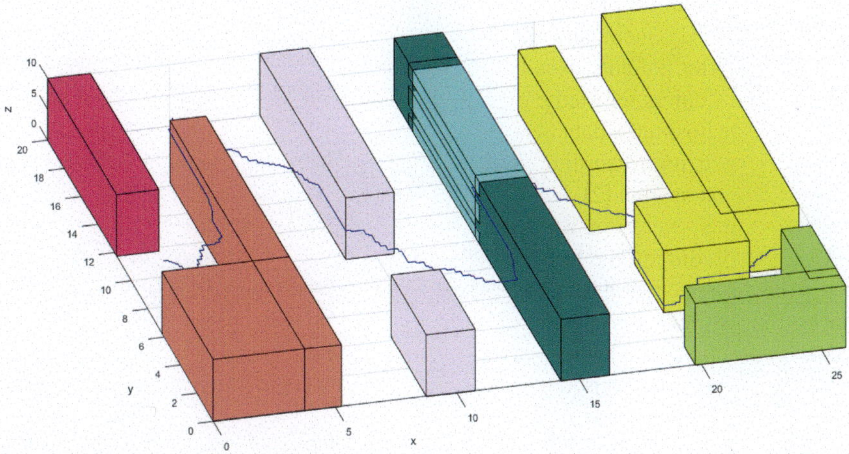

Fig. 3 Simulation of SSA in 3D static environment

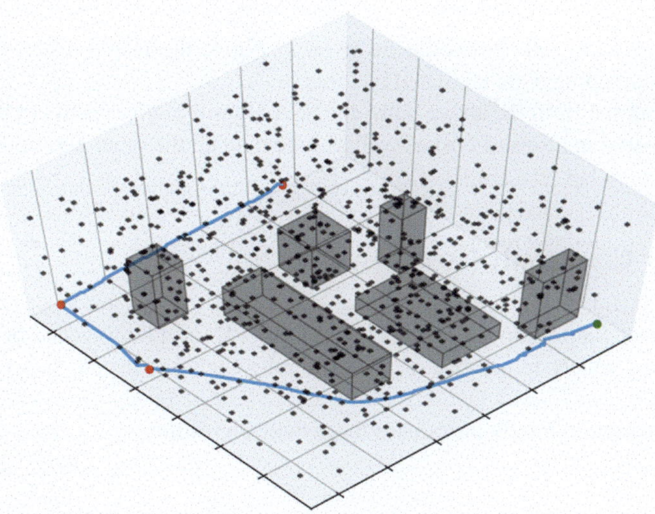

Fig. 4 Simulation of SSA in 3D dynamic environment [18]

4.5 Comparison of Results

To compare the results obtained by SSA, several other algorithms were also tested on the same static and dynamic 3D environments: Particle Swarm Optimisation (PSO, described in Sect. 5.1), Glowworm Swarm Optimisation (GSO) and Firefly Algorithm (FA, described in Sect. 5.3). More details about the experimental setup can be found in [19, 20].

Table 1 Results in static 3D environment

Algorithm	Population	Iteration	Best cost	Time
PSO	20	25	8.32×10^2	7.16×10^1
GSO	20	25	9.67×10^2	7.39×10^1
SSA	20	25	7.86×10^2	7.09×10^1

Table 2 Results in dynamic 3D environment

Algorithm	Population	Iteration	Best cost	Time
PSO	20	25	587468	588
GSO	20	25	561287	536
SSA	20	25	548139	271

For the static 3D environment given in Fig. 3, results obtained are tabulated in Table 1.

For the dynamic 3D environment given in Fig. 4, the results when the above algorithms are applied are tabulated in Table 2.

As we can see from the above table, the SSA algorithm gives better performance when compared in terms of cost as well as time than all the other algorithms.

5 Other Commonly Used NIAs

Apart from the Salp Swarm Algorithm (SSA), other Swarm Intelligence-based NIAs are also used widely in practice. A few examples include *Particle Swarm Optimisation (PSO)*, *Ant Colony Optimisation (ACO)* and *Firefly Algorithm (FA)*, among others. This section briefly introduces these three algorithms.

5.1 Particle Swarm Optimisation

The PSO technique was proposed by Dr Kennedy and Dr Earhart in 1995. PSO algorithm uses the swarming or social behaviour of flocks of birds and school of fish, as inspiration for particle optimisation [21]. In this technique, each bird or fish is considered as a *particle*, and together the school or the flock acts as a *swarm*. As in the case of SSA, the particles communicate with each other by learning from their experiences and also update themselves by searching the given space and building their memory. PSO would find any route existing in the environment.

Researchers apply PSO to many optimisation areas. PSO has a unique and straightforward searching mechanism. It is computationally, efficient and relatively easy to implement. Briefly, there are four vectors required to represent a particle in high-dimensional space: the current position, the best position so far, the best position in the entire neighbourhood, and the velocity. The details of PSO are left as further study [15].

5.2 *Ant Colony Optimisation*

Ant Colony Optimisation (ACO), as the name suggests, aims to depict the behaviour of ants in their natural environment [1]. Ants form *colonies*, which is analogous to swarm in case of SSA. The ants communicate between each other, and this allows them to compute the shortest route. This corresponds to the route between their nest (hive) and food source. Ants deposit a chemical substance called *pheromone* along their route. More ants follow the same route, the quantity of pheromone deposited grows and hence other ants can check out the higher pheromone-containing route. Pheromone evaporates with time, so only the most popular routes remain after some time.

As a result, the ACO algorithm works best when the source and destination of the problem are specific and clearly defined. The main steps for the ACO algorithm are as follows:

1. We generate and initialise the ants.
2. Iterate for each ant (break when we find the goal, or a stopping criteria is met).
3. Accumulate pheromone deposit along the visited route.
4. Daemon activities.
5. Evaporate the pheromone content of less popular route after some time.

5.3 *Firefly Algorithm*

The firefly algorithm was given by Xin-She Yang in 2008 [22]. It finds the global solution of an optimisation problem by swarm intelligence based on the *flashing behaviour* of fireflies. The firefly's flash operates as a signal to communicate with and warn other fireflies.

Similar to other algorithms of its kind, FA begins by generating random initial populations that can have feasible candidate solutions. The aim is that knowledge is collectively shared among all fireflies in the population. This guides the search for the best solution in the search space. Each particle has the capability to move in a multi-dimensional space and an attractiveness that is actively updated based on the knowledge of other fireflies and neighbours [23].

6 The Motivation from Nature

Now that we have given insights into some of the standard Nature-Inspired Algorithms, it is the right time to discuss the rationale behind using them for optimisation problems. NIAs are often more complex and time taking than conventional algorithms. Moreover, most NIAs are often less intuitive or require a great deal of research before devising the mathematical model to use them correctly.

However, NIAs are being used rapidly in many optimisation problems. We as humans love to explore and exploit the strengths of nature in almost every domain. A similar act in this domain can help researchers to make the best out of NIA and identify potential solutions for complex real-world scenarios. Moreover, to discover new methods and to improve the previous, the potential merits of interdisciplinary research are highlighted.

In the following sections, we dive a little deeper into the process of evolution in nature and related consider processes. Further, we describe the importance of metaphor in algorithm research along with, the way nature attains and provides encouragement for creativity. This section focuses on the aspects that we recognise as interesting and relevant, but it should not be considered as the only exhaustive resource which nature provides us.

6.1 Natural Selection and Optimisation

Darwin's theory is quite often quoted as, one of, if not the most important scientific theory in the nineteenth century. This can be attributed to the fact that this theory uses relatively simple rules of *reproduction*, *mutation* and *selection*, and combines then in a powerful way.

According to this theory, the fittest animals are ones who have superior problem-solving abilities. These abilities are the source of *strong* inspiration for many popular meta-heuristic-based optimisation algorithms. Although it is hard to say precisely what is being optimised by natural selection, we do observe it has produced certain useful features—*adaptation*, *efficiency* and *generality*. A detailed analysis of these features is left as further study [24].

6.2 Complex Systems and Emergent Behaviour

The presence of high-level function that develops due to the interaction among elements is termed as emergent behaviour. If the behaviour of the whole and not merely of its parts *(i.e., no-linear aggregate behaviour)*, then it is said to be a case of emergent behaviour. This, however, does not rule out a deterministic relationship between the parts and the whole.

Nature has the advantage of trial and error—a few hundred million years as such—and it enables nature for producing useful emergent behaviour by devising some rules. The rules often can be a student by an algorithm designer—with the possible help of a biological researcher—and be extracted into mathematical models, useful for generating computer algorithms.

In this way, algorithms that are simple enough to code can be generated from relatively complex processes. These emergent approaches to optimisation are also well suited to parallel computation, and as such, capable of utilising recent developments in the field.

6.3 Natural Metaphors

We, humans, tend to search for patterns in every facade of life. Researchers have found similar patterns between domain-specific problems and those which exist in nature. This has made us realise how nature solves a particular problem and harness that knowledge. Additionally, this *natural metaphor* enables us to think abstractly about different solutions.

7 Summary

It may come as no surprise that NIAs have a wide variety of applications in real-life scenarios. There has been tremendous growth in the domain of nature-inspired science in recent decades. These algorithms are often used for investigating nature by using computer simulations. This chapter provided a general walk-through of a nature-inspired algorithm as well as gives their detailed analysis and implementation on route planning problems. We hope that more researchers would take up this topic in the future, and the development of new and upgraded algorithms to solve fundamental human problems would be more in commonplace.

In conclusion, we would like to quote David Perkins in *Archimedes' Bathtub* by saying [25]:

> Mother Nature may re-purpose, but do we see the full pattern of breakthrough thinking in nature – the long search, little apparent progress, the precipitating event, some non-mental equivalent of the cognitive snap, and transformation? Arguably, yes!

References

1. T.T. Mac et al., Heuristic approaches in robot path planning: A survey. Robot. Autonom. Syst. **86**, 13–28 (2016)
2. J. Rosell, P. Iniguez, Path planning using harmonic functions and probabilistic cell decomposition, in *Proceedings of the 2005 IEEE International Conference on Robotics and Automation* (IEEE, 2005)
3. F.A. Cosio, M.A. Padilla Castaneda, Autonomous robot navigation using adaptive potential fields. Math. Comput. Model. **40**(9–10), 1141–1156 (2004)
4. N.N. Singh, A two-layered subgoal based mobile robot navigation algorithm with vision system and IR sensors. Measurement **44**(4), 620–641 (2011)
5. J. Lee, O. Kwon, L. Zhang, S.E. Yoon, A selective retraction-based RRT planner for various environments. IEEE Trans. Robot. **30**(4), 1002–1011 (2014)
6. B. Lau, C. Sprunk, W. Burgard, Efficient grid-based spatial representations for robot navigation in dynamic environments. Robot. Autonom. Syst. **61**(10), 1116–1130 (2013)
7. B. Park, J. Choi, W.K. Chung, An efficient mobile robot path planning using hierarchical roadmap representation in indoor environment, in *2012 IEEE International Conference on Robotics and Automation* (IEEE, 2012)
8. V.R. Desaraju, J.P. How, Decentralized path planning for multi-agent teams in complex environments using rapidly-exploring random trees, in *2011 IEEE International Conference on Robotics and Automation* (IEEE, 2011)
9. A.-M. Zou et al., Neural networks for mobile robot navigation: a survey, in *International Symposium on Neural Networks* (Springer, Berlin, 2006)
10. H. Chang, T. Jin, *Command Fusion Based Fuzzy Controller Design for Moving Obstacle Avoidance of Mobile Robot*. Future Information Communication Technology and Applications (Springer, Dordrecht, 2013), pp. 905–913
11. N.S. Pal, S. Sharma, Robot path planning using swarm intelligence: a survey. Int. J. Comput. Appl. **83**(12), 5–12 (2013)
12. H. Zang, S. Zhang, K. Hapeshi, A review of nature-inspired algorithms. J. Bion. Eng. **7**, S232–S237 (2010)
13. P. Agarwal, S. Mehta, Nature-inspired algorithms: state-of-art, problems and prospects. Int. J. Comput. Appl. **100**(14), 14–21 (2014)
14. A. Parashar, K.K. Swankar, Genetic algorithm using to the solution of unit commitment. Int. J. Eng. Trends Technol. **4**(7), 2986–2990 (2013)
15. S. Binitha, S. Siva Sathya, A survey of bio inspired optimization algorithms. Int. J. Soft Comput. Eng. **2**(2), 137–151 (2012)
16. S. Mirjalili et al., Salp swarm algorithm: a bio-inspired optimizer for engineering design problems. Adv. Eng. Softw. **114**, 163–191 (2017)
17. P.A.V. Anderson, Q. Bone, Communication between individuals in salp chains. II. Physiology. Proc. R. Soc. Lond. Ser. B. Biol. Sci. **210**(1181), 559–574 (1980)
18. M.D. Solomon, A development of a real-time hierarchical 3D path planning algorithm for unmanned aerial vehicles, https://github.com/mds1/path-planning/tree/master/paper
19. P. Pandey, A. Shukla, R. Tiwari, Three-dimensional path planning for unmanned aerial vehicles using glowworm swarm optimization algorithm. Int. J. Syst. Assuran. Eng. Manag. **9**(4), 836–852 (2018)
20. R.K. Dewangan, A. Shukla, W. Wilfred Godfrey, Three Dimensional path planning using Grey wolf optimizer for UAVs. Appl. Intell. **49**(6), 2201–2221 (2019)
21. H.I. Kang, B. Lee, K. Kim, Path planning algorithm using the particle swarm optimization and the improved Dijkstra algorithm, in *2008 IEEE Pacific-Asia Workshop on Computational Intelligence and Industrial Application*, vol. 2 (IEEE, 2008)
22. X.-S. Yang, *Nature-Inspired Metaheuristic Algorithms* (Luniver Press, 2010)

23. D.I. Esa, A. Yousif, Scheduling jobs on cloud computing using firefly algorithm. Int. J. Grid Distrib. Comput. **9**(7), 149–158 (2016)
24. K.C.B. Steer, A. Wirth, S.K. Halgamuge, The rationale behind seeking inspiration from nature, in *Nature-Inspired Algorithms for Optimisation* (Springer, Berlin, 2009), pp. 51–76
25. D.N. Perkins, *Archimedes' Bathtub: The Art and Logic of Breakthrough Thinking* (W.W. Norton, 2000). ISBN 10.9780393047950

Statistical Time Series Models for Wind Speed Forecasting

Anil Kumar Kushwah, Rajesh Wadhvani, and Varsha Kushwah

Abstract This work was developed for forecasting wind speed data by using various statistical models, which can further be utilized for the estimation of the annual energy production of commercial wind farms. The pattern of historical data available for modeling may be linear or nonlinear. For linear time series pattern, Autoregressive Integrated Moving Average (ARIMA), ARIMAX model by using exogenous variable X and Vector Autoregressive (VAR) models have been developed. In order to achieve good performance on the nonlinear pattern, Generalized Autoregressive Score (GAS), GAS with exogenous variable (GASX) model has been developed. Wind time series data taken from different site locations of the National Renewable Energy Laboratory (NREL) repository have been utilized to develop the models. In our investigation, it has been found that the VAR model performs the best in most of the cases since it includes more than one variable for the development of the model.

1 Introduction

The wind energy is the most utilized energy source, which is getting more attention in the energy production society. The use of wind energy will not pollute the environment because it is a natural and clean source of energy, and it shrinks the consumption of traditional energy assets. Wind energy is suitable for a system that requires continuous energy [1]. Wind speed is the main factor of wind energy which is fluctuating as well as nonstationary in nature. The accurate prediction of wind speed is very important for the effective utilization of wind power generation. The

A. Kumar Kushwah (✉) · R. Wadhvani · V. Kushwah
Department of Computer Science and Engineering, Maulana Azad National Institute of Technology, Bhopal 462003, Madhya Pradesh, India
e-mail: akushwah190@gmail.com

R. Wadhvani
e-mail: rajeshwadhvani@manit.ac.in

V. Kushwah
e-mail: vk8415@gmail.com

A. Nanda and N. Chaurasia (eds.), *High Performance Vision Intelligence*, Studies in Computational Intelligence 913,
https://doi.org/10.1007/978-981-15-6844-2_16

main aim to develop a model for wind speed prediction is based on given wind time series data. Therefore, wind speed prediction is a famous research area because it is utilized in wind power generation, which produces the wind energy [2].

In recent years, various researchers are doing the study of wind speed forecasting research areas. According to the time horizon, wind speed prediction can be classified into three categories such as long-term forecasting, medium-term forecasting, and short-term forecasting [3]. Wind speed forecasting models are divided into three broad categories, namely, physical models, statistical models, and artificial intelligence models [4]. Physical models are generally used in long-term prediction, which is based on numerical weather prediction [5]. However, the mathematical calculation is very complex and takes more time that why it uses rarely. On the other hand, statistical models are forecast based on historical data by maintaining the relationship between the dependent and independent variables. The statistical models like Autoregressive (AR), Moving Average (MA), Autoregressive Moving Average (ARMA) [6], and Autoregressive Integrated Moving Average (ARIMA) [7] used the past data to obtain the seasonal and trend components of wind time series data. Moreover, there are diverse factors that affect the wind speed series and that require complex function in order to capture relationships between variables. Although, for wind speed forecasting several artificial intelligence models are used, namely, Support Vector Machine (SVM) [8], Artificial Neural Network (ANN) [9–11], and Extreme Learning Machine (ELM) [12]. These are capable of handling the nonlinear relationship present in the time series data.

In brief, many statistical methods are used for wind power forecasting, Rajagopalan and Santoso [13] have proposed the ARMA model for prediction of wind speed data which is used in forecasting the time series data in one hour ahead of the time scale. Buyuksahn and Ertekin et al. [14] have suggested the ARIMA model is used for Wolf's sunspot nonstationary time series data. That model gives the result compared with persistence models. Camelo and Lucio et al. [15] have proposed that ARIMAX model is used for wind power time series meteorological data like wind speed, temperature, pressure with exogenous factor. Kalli and Griffin et al. [16] used the VAR model which gives the proper way of generating the dynamic interactions of multiple time series data. It is mainly used in the macroeconomic analysis. Kushwah and Wadhvani [17] have applied the GAS and GASX model for wind speed forecasting on the 5-min time interval. They have found that the GASX model outperforms as compared to ARIMA and GAS model. The accuracy of all statistical models are measured by using Mean Absolute Error (MAE), and Root Mean Squared Error (RMSE).

This chapter work is ordered as follows. In Sect. 2, it is defined the different types of statistical models for wind speed forecasting. Section 3 represented the curves of wind speed prediction for all of these statistical models, analysis of all results, and accuracy. Section 4 conclusion of this work.

2 Statistical Models for Wind Speed Forecasting

2.1 Autoregressive Moving Average (ARIMA) Model

Buyuksahin and Ertekin [14] have proposed ARIMA model. The ARIMA is generally used in linear time series data which means the future value from a variable is predicted by the past observations using the linear function. ARIMA is expected that time series data is present in the stationary form. ARIMA checks if the given time series is stationary or not. For checking the stationarity of data, the value of mean, variance, and autocorrelation are constant over time. If data does not follow these properties, then data is nonstationary. Nonstationary data is converted into stationary time series data by applying the differencing method of ARIMA until data is not converted into stationary data. ARIMA model contains AR model (order p), MA model (order q), and d is the number of differentiation terms. The representation of ARIMA model is defined as

$$y_t = \delta + \sum_{i=1}^{p} \phi_i y_{t-i} + \sum_{j=1}^{q} \theta_j \epsilon_{t-j} + \epsilon_t \tag{2.1}$$

where y_t is the time series value at time t and ϵ_t is the error term at time t. δ, ϕ_i, and θ_j are the model parameters.

2.2 Autoregressive Moving Average with Exogenous Variable (ARIMAX) Model

Camelo and Lucio et al. [15] have proposed the ARIMAX model as an extension of the ARIMA model through the inclusion of exogenous variable X. ARIMAX can be used to model multivariate time series data. The notation ARIMAX (p, d, q, r) refers to the ARIMAX model with p autoregressive terms, q moving-average terms, d is the degree of the differencing term, and r is the number of exogenous variables. This model can be described as:

$$y_t = \rho + \sum_{i=1}^{p} \beta_i y_{t-i} + \sum_{j=1}^{r} \omega_j w_j + \sum_{k=1}^{q} \theta_k \epsilon_{t-k} + \epsilon_t \tag{2.2}$$

where ρ is the constant, y_{t-i} is a dependent variable, β_i is a coefficient of y_{t-i}, w_j represents the exogenous variables, ω_j represents the coefficient of exogenous variables, ϵ_{t-k} is the random error, θ_k is the coefficient of the ϵ_{t-k}, and ϵ_t represents the error of the ARIMAX model.

2.3 Vector Autoregressive (VAR) Model

Skripnikov et al. [18] have proposed Vector Autoregressive model (VAR). The VAR model is generally used for multivariate time series data for forecasting the time series. It is a generalized form of a univariate autoregressive model which is used in dynamic multivariate time series data. In the VAR model, all the variables behave symmetrically, and in a mathematical equation, it is evolved based on its legs and the legs of all the other variables in the model is defined by each variable of the equation. In the VAR model, parameters are generally nominal in a batch manner, and the noise terms are assumed to be Gaussian. The basic equation of the VAR model that uses the lag order of D is defined as

$$X_t^k = A_1^k X_{t-1}^k + \cdots + A_D^k X_{t-D}^k + \epsilon_t^k, \epsilon_t^k \sim N(0, \sigma_k^2 I_p) \tag{2.3}$$

where ϵ_t^k is white noise, A_d^k is a $p * p$ transition matrix that effects of order the p variables for multidimensional k, $X_t^k = (X_{1,t}^k, ..., X_{2,t}^k)^T$ is a p variate stationary time series, t = D,...,T, d = 1,...,D , and k = 1,...,K. However, we suppose the diagonal noise covariance matrix is $\sum_k = \sigma_k^2 I_p$, which is allowed to divide a problem into subproblems that is why it is solved in parallel.

2.4 Generalized Autoregressive Score (GAS) Model

Generalized Autoregressive Score (GAS) models are observation-driven time series data model constructed by the score function for nonlinear data. Advantage of the GAS model is observation-driven model because it is an extension to long-term data, asymmetric data, and other complicated dynamics data can be introducing without complexity. For a conditional observation density $p(y_t|f_t)$ with an observation y_t and a latent time-varying parameter f_t, we assume the parameter f_t follows the given equation:

$$f_t = \mu + \sum_{i=1}^{p} \phi_i f_{t-i} + \sum_{j=1}^{q} \alpha_j S(f_{j-1}) \frac{\partial logp(y_{t-j}|f_{t-j})}{\partial f_{t-j}} \tag{2.4}$$

where ϕ is the autoregressive coefficient, μ is the intercept, α is the scaling parameter, S is the positive scaling parameter, p and q are the order of the GAS model. S is multiplied with the first derivative of $p(y_t f_t)$ that contribution for one observation at time j.

2.5 Generalized Autoregressive Score with Exogenous Variable (GASX) Model

The GAS model with exogenous variable (GASX) model is an extended form of the GAS model to implement with exogenous variable X. There is conditional observation density $p(y_t f_t)$ that is found by using observation y_t and time-varying parameter f_t. Where f_t is followed the equation as

$$f_t = \mu + \sum_{k=1}^{K} \beta_k X_{t,k} + \sum_{i=1}^{p} \phi_i f_{t-i} + \sum_{j=1}^{q} \alpha_j S(f_{j-1}) \frac{\partial logp(y_{t-j} | f_{t-j})}{\partial f_{t-j}} \tag{2.5}$$

where ϕ is the autoregressive coefficient, μ is the intercept, α is the scaling parameter, X is the exogenous variable, β is the coefficient, S is the positive scaling parameter, k, p, and q are the order of GASX model. S is the positive scaling parameter, p and q are the order of the GAS model. S is multiplied with the first derivative of $p(y_t f_t)$ that is the contribution for one observation at time j. The main advantage of GASX model is used as the additional factor in the modeling to boost the accuracy of model.

3 Experiments and Discussions

3.1 Dataset

Here we are going to show the implementation of statistical models which are discussed in previous sections. The datasets are received from NREL (National Renewable Energy Laboratory) with different site id such as 124693, 69015, 16883, 36363, 44402, and 45208. The site id 12469 is having a geographical location with longitude −120.005° and latitude 46.9016° which has an average wind speed of 6.744 m/s. The site id 69015 is having a geographical location with longitude −115.13° and latitude 40.267°, which has an average wind speed of 7.623 m/s. The site id 45208 is having a geographical location with longitude −104.695° and latitude 39.152° which has 7.214 m/s as average wind speed. NREL has 5 min on an average wind speed values received from SCADA. The wind speed is recorded at 100 m height in this system. The dataset has 105120 observations that are noted from January 2012 to December 2012.

3.2 Measuring Criteria

statistical models have been developed which show the behavior of the actual data, the determination of appropriate criteria to evaluate the capacity of a model, to sum up, it is likewise significant. In our experimentation, Mean Absolute Error (MAE) and Root Mean Squared Error (RMSE) measuring criteria are used to test the performance of wind speed forecasting. The MAE and RMSE are defined as follows:

$$MAE = \frac{1}{N}\sum_{i=1}^{N}(y_i - x_i) \tag{3.1}$$

$$RMSE = \sqrt{\frac{1}{N}\sum_{i=1}^{N}(y_i - x_i)^2} \tag{3.2}$$

where N is the total number of data points, y is the forecasted variable, and x is the input variable. The model which gives the lowest values of MAE and RMSE performs better. All the statistical models which are discussed in the previous section are implemented in Python 3.6 version.

3.3 Results Analysis

For our experimentation, all the models are applied for wind speed forecasting on six different datasets that are available as 5 min interval. The performance of the statistical models is measured by MAE and RMSE values. The model which gives the minimum values of MAE and RMSE performs better. In this research work, ARIMA, ARIMAX, VAR, GAS, and GASX models are used for wind speed forecasting. The wind speed data that taken from NREL are divided into training and testing pair. The training data contains 75% of actual data, whereas testing data contains 25% of actual data. The predicted and actual wind speed is shown in Fig. 1 using the ARIMA model on dataset 124693. Figure 2 shows the actual and predicted wind speed using the ARIMAX model on dataset 124693. It has been observed that the ARIMAX model performs better in terms of accuracy as compared to the ARIMA model.

Similarly, Figs. 3, 4, and 5 show the actual and predicted wind speed using the VAR, GAS, and GASX model, respectively, on the 124693. The VAR model shows better accuracy as compared to other models on dataset 124693.

Figure 6 demonstrates the original and predicted wind speed using the ARIMA model on dataset 36363. Figure 7 shows the original and predicted wind speed using the ARIMAX model on dataset 36363. It has been observed that the ARIMAX model performs better in terms of accuracy as compared to the ARIMA model because it includes the temperature as an exogenous variable.

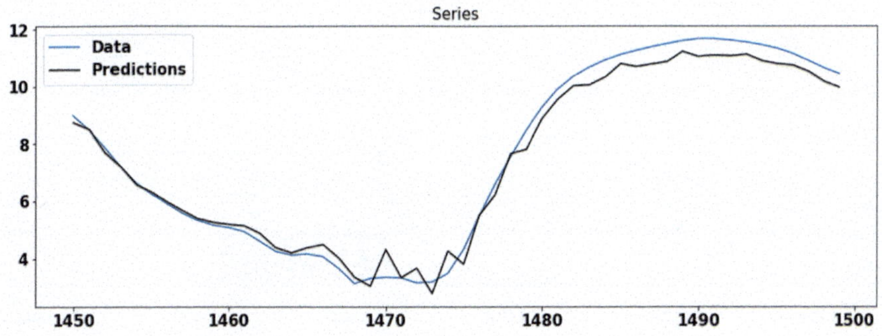

Fig. 1 Wind speed forecasting using ARIMA model for dataset 124693

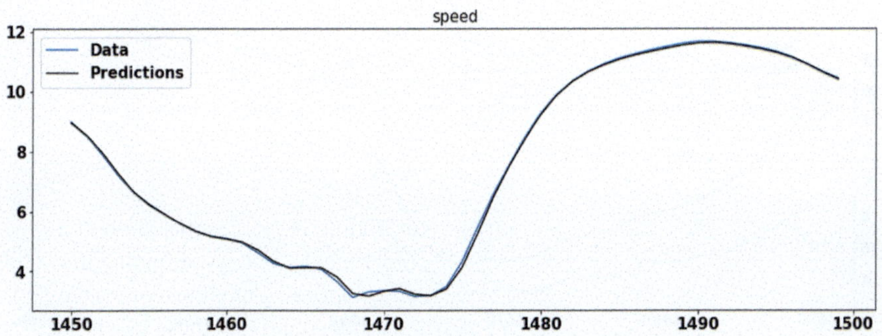

Fig. 2 Wind speed forecasting using ARIMAX model for dataset 124693

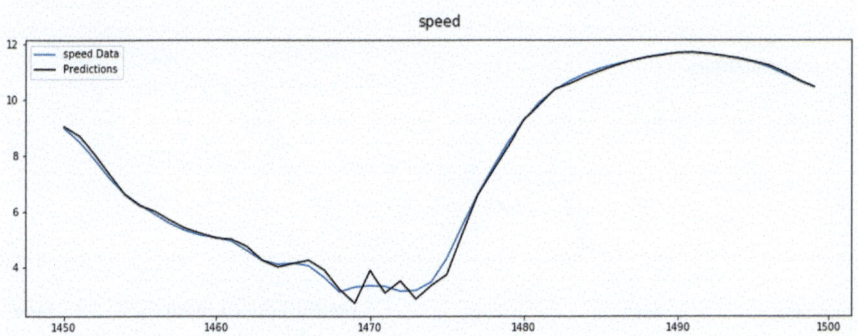

Fig. 3 Wind speed forecasting using VAR model for dataset 124693

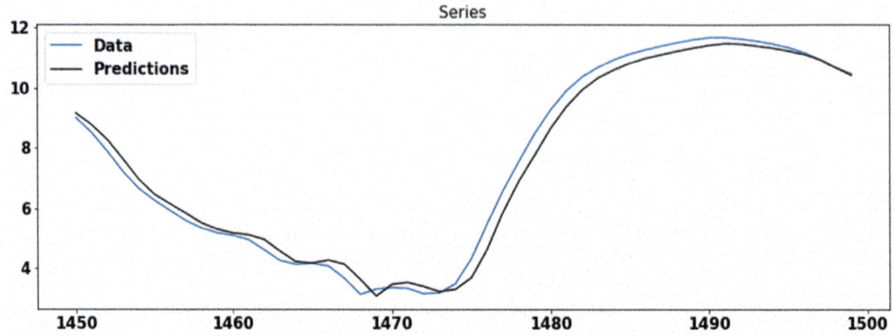

Fig. 4 Wind speed forecasting using GAS model for dataset 124693

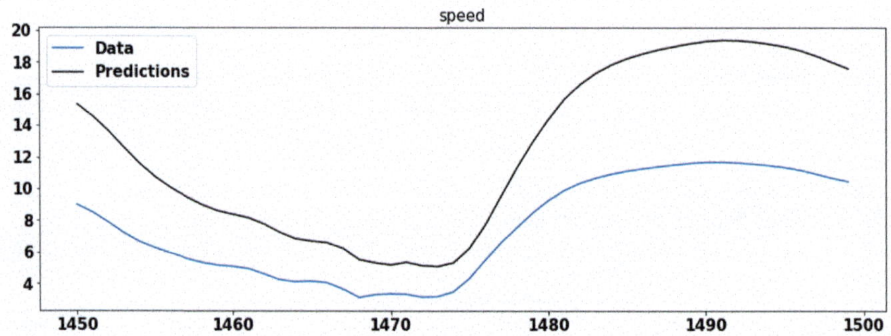

Fig. 5 Wind speed forecasting using GASX model for dataset 124693

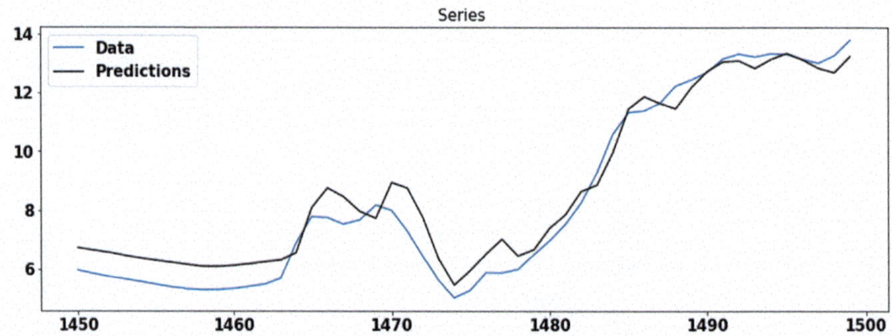

Fig. 6 Wind speed forecasting using ARIMA model for dataset 36363

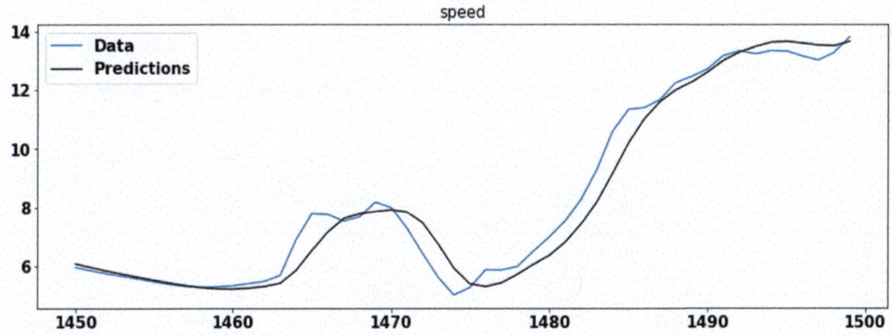

Fig. 7 Wind speed forecasting using ARIMAX model for dataset 36363

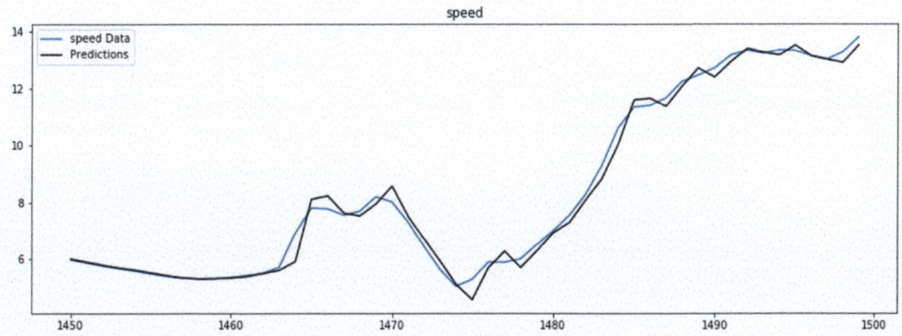

Fig. 8 Wind speed forecasting using VAR model for dataset 124693

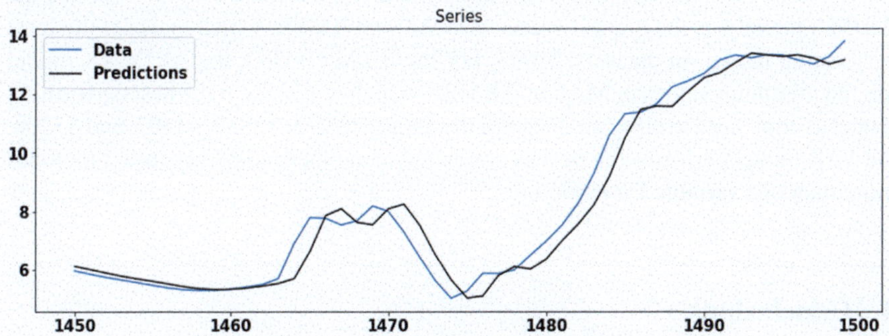

Fig. 9 Wind speed forecasting using GAS model for dataset 124693

Similarly, Figs. 8, 9, and 10 show the actual and predicted wind speed using the VAR, GAS, and GASX model, respectively, on the dataset 36363. The ARIMAX model shows better accuracy as compared to other models on dataset 36363.

Table 1 demonstrates the estimated values of MAE and RMSE using the ARIMA, ARIMAX, VAR, GAS, and GASX models on different datasets. In Table 1, the VAR model has minimum MAE and RMSE values as compared to other models, so we can

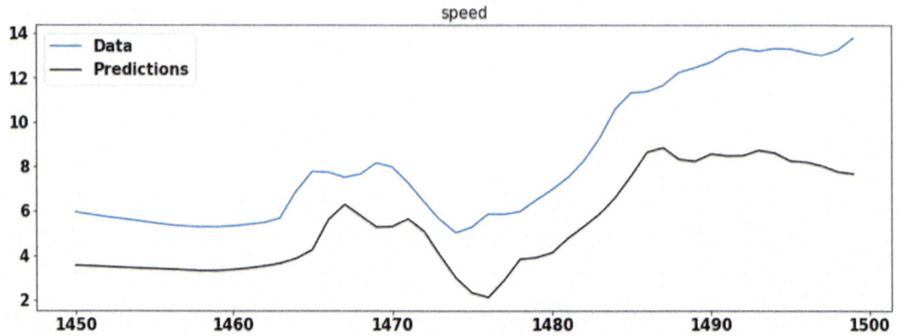

Fig. 10 Wind speed forecasting using GASX model for dataset 124693

Table 1 Estimated values of MAE and RMSE on different dataset

Dataset	ARIMA	ARIMAX	VAR	GAS	GASX
124693	11.419/12.447	3.858/4.503	2.924/3.375	3.959/4.234	3.994/4.655
69015	3.322/3.969	4.924/5.679	3.104/4.461	8.237/9.598	4.783/5.573
16883	6.765/7.248	5.389/6.104	2.866/3.641	4.966/5.588	4.283/5.202
36363	6.223/7.022	2.858/3.255	3.002/4.251	5.973/6.776	3.032/3.471
44402	4.735/5.343	7.345/8.688	4.494/6.124	5.022/6.328	5.815/5.041
45208	4.169/5.387	4.365/5.519	4.082/4.889	4.185/5.441	4.533/5.766

say the VAR model is the best performing model on dataset 124693. Whereas, the ARIMA model has the highest values of MAE and RMSE, which makes it the least performing model on dataset 124693. For the dataset 36363, the ARIMAX model has the minimum values of MAE and RMSE, which will make it the best performing model as compared to other models. For the dataset 69015, 16883, 44402, and 45208, the VAR model performs the best as compared to another model because it includes more than one variable for modeling.

4 Conclusions

Accurate wind speed forecasting plays a crucial role in wind energy generation. In this work, several statistical models are investigated on different datasets that are taken from NREL. In the experimentations, it has been found that the VAR model performs well most of the time on different datasets. In our experimentation, the ARIMAX model performs well only for dataset 36363. Whereas, the VAR model performs best for all the remaining datasets. All the statistical models were evaluated by using the MAE and RMSE values. In the future, the hybrid models can be offered as a forecasting modeling approach which will show consistent performance on the different datasets.

References

1. L. Xie, P.M.S. Carvalho, L.F.M. Ferreira, Liu Juhua, B.H. Krogh, N. Popli, Wind integration in power systems: operational challenges and possible solu-tions. Proc. IEEE **99**, 214–232 (2011)
2. J. Jung, R.P. Broadwater, Current status and future advances for wind speed and power forecasting. Renew. Sustain. Energy Rev. **31**, 762–777 (2014)
3. J. Yan, Y. Liu, S. Han, Y. Wang, S. Feng, Reviews on uncertainty analysis of wind power forecasting. Renew. Sustain. Energy Rev. **52**, 1322–1330 (2015)
4. H. Liu, Tian Hqi, Li Yfei, Comparison of two new ARIMA-ANN and ARIMA Kalman hybrid methods for wind speed prediction. Appl. Energy **98**, 415–424 (2012)
5. J. Zhao, Z.H. Guo, Z.Y. Su, Z.Y. Zhao, X. Xiao, F. Liu, An improved multi-step forecasting model based on WRF ensembles and creative fuzzy systems for wind speed. Appl. Energy **162**, 808–826 (2016)
6. D. Yang, V. Sharma, Z. Ye, L.I. Lim, L. Zhao, A.W. Aryaputera, Forecasting of global horizontal irradiance by exponential smoothing, using decompositions. Energy **81**, 111–119 (2015)
7. J.L. Torres, A. García, M. De Blas, A. De Francisco, Forecast of hourly average wind speed with ARMA models in Navarre (Spain). Solar Energy **79**, 65–77 (2005)
8. J. Hu, J. Wang, K. Ma, A hybrid technique for short-term wind speed prediction. Energy **81**, 563–574 (2015)
9. J. Koo, G.D. Han, H.J. Choi, J.H. Shim, Wind-speed prediction and analysis based on geological and distance variables using an artificial neural network: a case study in South Korea. Energy **93**, 1296–1302 (2015)
10. T. Fu, C. Wang, A hybrid wind speed forecasting method and wind energy resource analysis based on a swarm intelligence optimization algorithm and an artificial intelligence model. Sustainability **10**(11), 1–24 (2018)
11. D. Liu, J. Wang, H. Wang, Short-term wind speed forecasting based on spectral clustering and optimized echo state networks. Renew Energy **78**, 599–608 (2015)
12. V. Nikolic, S. Motamedi, S. Shamshirband, D. Petkovic, S. Ch, M. Arif, Extreme learning machine approach for sensorless wind speed estimation. Mechatronics **34**, 78–83 (2015)
13. S. Rajagopalan, S. Santoso, Wind power forecasting and error analysis using the autoregressive moving average modeling. In: IEEE Power Energy Society General Meeting, pp. 1–6 (2009)
14. U.C. Buyuksahin, S. Ertekin Improving forecasting accuracy of time series data using a new ARIMA-ANN hybrid method and empirical mode decomposition. Neurocomputing (2019)
15. H. do Nascimento Camelo, P.S. Lucio, J.B. Vercosa L. Junior, P.C. Marques de Carvalho, D. von Glehn dos Santos, Innovative hybrid models for forecasting time series applied in wind generation based on the combination of time series models with artificial neural networks. Energy **151**, 347–357 (2018)
16. Maria Kalli, Jim E. Griffin, Bayesian nonparametric vector autoregressive models. J. Econom. **203**, 267–282 (2018)
17. A.K. Kushwah, R. Wadhvani, Performance Monitoring of Wind Turbine Using Advanced Statistical Methods. Sadhana, Springer, vol. 44(163) (2019)
18. A. Skripnikov, G. Michailidis, Regularized joint estimation of related vector autoregressive models. Comput. Stat. Data Anal. **139**, 164–177 (2019)

Smart HealthCare Model: An End-to-end Framework for Disease Prediction and Recommendation of Drugs and Hospitals

Megha Rathi, Nimit Jain, Priya Bist, and Tarushi Agrawal

Abstract The need for an intelligent healthcare system is indispensable. The limited knowledge and experience of doctors lead to inaccurate predictions and diagnosis of the health-related issue. Patients also have very little knowledge about the drugs that have been prescribed to them. This kind of knowledge is crucial for the patients to judge their diagnosis done by the doctor. To make people familiar with this, we have proposed a model. This model consists of three phases. (1) The first phase deals with the disease prediction on the basis of symptoms entered by the patient. (2) The second phase will suggest the patient the best drug suitable for his condition. (3) The last phase is recommending the best hospital for his/her treatment. Various machine learning algorithms have been applied to the dataset. Results show that the model is able to give medical guidance precisely and effectively.

1 Introduction

Health has become a concern for the new generation due to changes in the life cycle of a person. The risk for health diseases has increased in the past thus the searches for disease according to symptoms have increased. According to a survey by Google 1 in every 20 searches is about health. Medical errors have led to the death of more than 200 thousand people in China and around 100 in the US. In 2017 summer,

M. Rathi (✉) · N. Jain · P. Bist · T. Agrawal
Computer Science & Engineering Department, Jaypee Institute of Information Technology, Noida, India
e-mail: megha.rathi@jiit.ac.in

N. Jain
e-mail: j.nimit99@gmail.com

P. Bist
e-mail: priyagopalbist9@gmail.com

T. Agrawal
e-mail: tarushi.agrawal17@gmail.com

A. Nanda and N. Chaurasia (eds.), *High Performance Vision Intelligence*, Studies in Computational Intelligence 913,
https://doi.org/10.1007/978-981-15-6844-2_17

dot-health conducted a survey of 1,509 online respondents to know the portion of health-related searches by them [1]. This is also because of the increase in the health issues in today's world. The number of errors regarding the healthcare system is also at a hype.

The main reason for the increase in the online searches for the health-related issue is due to medical mafia. The contemporary doctors try to prescribe unnecessary drugs and treatment to charge from patients a great lump sum amount of money. Some of the reasons for this increase are: (1) People try to self-diagnose their medical issues before going to the doctor as it is really an expensive affair, (2) Some people even try to check the remedy suggested by professionals due to increase in the number of wrong medical treatment, (3) It helps in better recommending of hospitals on the basis of disease and doctors. Sometimes there is also a need to recommend a second doctor as the previous medical solution was not able to satisfy the patient, and (4) there are times when a doctor recommends the wrong drug due to the limited knowledge that they have from their experience.

It is very important to choose the best healthcare institution for specific health related to the problem. These institutes are selected on the basis of track records of health quality care provided and the desirable and undesirable outcomes in the past. The effectiveness of a hospital depends on the service provided to the patients, patient care management, and patient satisfaction. Physicians, one who recommend the hospitals to the patients, usually have knowledge about the facilities which provide the best care. Centers for Medicare and Medicaid Services are putting a lot of efforts in making the hospital's data available to the physician and to the general public to assist their decision. According to a research 2600 lives could be saved by improved hospital referral system [2].

In this paper, we are introducing the solution for the problem of detecting the disease and predicting the best medicine with the technology of machine learning. The paper also studies the recommendation of hospitals according to the location and specific disease it is specialized in. First, the symptoms of the patient are taken into consideration. The symptoms are analyzed and disease is predicted by training the dataset using a supervised learning algorithm [3]. The best drug is then identified from the list of medicines for the specific disease. In the whole designing of the work, the machine learning technology has been used. The structure of the paper is basically divided into three models: the disease prediction, drug prediction, and hospital recommendation

2 Literature Review

Earlier patient's health records were stored in the form of Electronic Health Records. Various data mining and pattern recognition techniques were used to extract valuable information from these records. A method called Evidence-Based Medicine (EBM) was suggested [4]. In EBM decisions related to health were taken on the basis of current clinical evidence in addition to the doctor's experience. EBM has been applied in neurology and pediatric urology. Researchers have tried to support EBM by introducing a decision support system using big data analytics [5]. It provides enhanced access to care and enhanced clinical decision-making technique. The current medical technology used to predict the disease is not very efficient, leading to treatment delays or irrelevant diagnoses which can turn out to be fatal. Some studies considered delays in the prediction of the diseases [6].

Lack of knowledge of people toward health symptoms is also a big issue which led to an increase in delay of treatment of disease like Tuberculosis. Location is also a barrier in the delay and this research has tried to focus on the factors that lead to delay of the health treatment. Disease risk prediction had also been done using big data to reduce problems faced due to incomplete data [7]. Moreover, this paper has focused on the disease risk prediction considering the location so that chronic disease risk prediction in disease frequent communities can be efficient. CNN algorithm had been suggested for the problem. The latent factor was used to overcome the problem of missing data. A survey was conducted to know the time interval between the onsets of symptoms to the first treatment. There is scope for advancement in medicine area using big data and cloud computing techniques. Research in resource management of cloud computing platforms, which has application in medical data mining has been highlighted [8].

A research had been done in the past to relate technology and medication using the adaptive web. The goal of the research was to provide best medical treatment to a consumer using the information of past experiences of the other users [9]. This work also points out the side effect that a consumer can go through if a particular medicine is prescribed to him/her. Web-based system had been proposed to scratch past healthcare information especially past experiences related to it. A medication recommender system had been proposed in the past to prescribe an anti-diabetic drug using domain Ontology and Semantic Web Rule Language (SWRL) [10]. Domain ontology is to gather review about a drug from a hospital specialist in a database of American Association of Clinical Endocrinologists Medical Guidance (AACEMG). Throughout the work, ontology knowledge had been generated using different attributes on drug nature. Semantic Web Rule language had also been used to accurately recommend a drug to a diabetic patient with least side effects.

Various studies have analyzed that the hospitals which participated in the JCAHO survey had higher quality and results [11]. Joint Commission on Accreditation of Healthcare Organizations (JCAHO) is a non-profit organization which gives hospitals accreditation. JCAHO accreditation is given to a hospital only if it meets certain standards. However, this accreditation has restricted use in distinguishing the separate performance of the accredited hospitals. A new method aimed at estimating the minimum standard volume of facilities for healthcare center and how long distance should he travel for the healthcare centers was devised [12]. Traveling will not be a burden for the patient if it is a high volume hospital and a small change in volume does not lead to change in the traveling burden.

Studies have shown that patients often prefer higher risk hospitals which are closer to them as compared to far lower risk hospitals [13]. As a result of which, a large number of deaths occur which were potentially avoidable. Geographic factors may influence the result of institutional predictors. Reference [14] predicted that the volume threshold is not a perfect solution for a largely rural state. The leafgroup focuses on evidence-based referral system using a threshold volume for a non-emergent situation. It analyzes whether a hospital is able to satisfy the threshold volume standard. Initial conclusions indicated that the distance of the hospital also plays a major role in the selection of healthcare institutes in rural areas, which has been neglected in the above research.

To know the review of a hospital, research had been conducted using the concepts of sentiment analysis [15]. It was proposed that sentiment analysis of online unstructured comments by the patient about their health care. The result was also compared with a survey. It was concluded that machine learning was able to predict efficiently the assessment of hospital by patients. Machine learning was used to find infection that patients suffered from due to a clinical test conducted in a specified hospital. The research was aimed to help customer rate the hospital [16]. Normal surveillance methods would have been time-consuming thus classification and clustering algorithm like SVM had been used for the purpose. It was concluded that the novel resampling strategy had been highly beneficial in the prediction.

Recommender systems aim to provide personalized products to the users based on their previous ratings or likeliness with other users. The recommendation is mostly used in areas such as online shopping, movie recommendation, or a book recommendation. Medicine includes rare recommendation technologies. Our proposed system aims to provide a complete healthcare system including disease diagnosis and drug and hospital recommendation.

3 System Design and Architecture

In this section the overall framework of the work is described along with the algorithm and dataset collection. The data is first preprocessed and is further used in the study. Additionally, the section describes the user interface and the procedure which led to the prediction of the disease. It also describes the recommendation technique used in the work to recommend the drugs and hospital based on the predicted disease. The main aim of the work is to propose a smart healthcare system which aids the patient in self-diagnosis and enables them to take an informed decision of their treatment.

3.1 Data Collection

Three datasets have been used in this work.

The first dataset is used for disease prediction. The dataset has been obtained from a kaggle repository predicting disease from using ml(). It consists of 3993 instances and 132 attributes. Symptoms are the attributes corresponding to the diseases. Each instance is a vector of 1 and 0, where 1 represents the presence of the symptom and 0 represents the absence of that symptom. There are in total 34 unique diseases [17] (Table 1).

The second dataset is the Drug Review dataset. The dataset has been obtained from UCI machine learning repository [Drug Review Dataset (Drugs.com) Data Set]. It consists of 44509 instances and 5 attributes. It contains the drug names for a specific disease along with the user review, rating, and useful count. Rating is the score, between 1 and 10, given by the majority of the users and the useful count is the count of the users who found the drug useful [18] (Table 2).

The third dataset is the Hospital Directory dataset. The dataset has been obtained from the Indian Government site (health.gov.in). There are 30273 instances and 48 attributes. Only 7 of them are appropriate for the model [19]. The attributes are listed in Table 3.

Table 1 Symptoms dataset

S. No.	Name	Description
1.	Itching	List of Symptoms [Binary]
2.	Skin rash	
3.	Nodal skin eruptions	
4.	Continuous sneezing	
5.	Shivering	
6.	Chills	
7.	Joint pain	
8.	Stomach pain	
9.	Acidity	
10.	Ulcers on tongue	
11.	Prognosis	Predicted disease

Table 2 Drug review dataset

S. No.	Name	Description
1.	Drug name	Name of the drug [character]
2.	Condition	Name of the disease [character]
3.	Review	Feedback of drug users in the past [character]
4.	Rating	Score given by the user in the range 1–10 [integer]
5.	Useful	Count number of users who found it useful [integer]

Table 3 Hospital directory dataset

S. No.	Name	Description
1.	Hospital name	Name of the hospital [character]
2.	Hospital category	Category of the hospital (Public/Private) [character]
3.	Discipline systems of medicine	Types of medicines prescribed (Allopathy/Homeopathy/Ayurveda) [character]
4.	Address original first line	The first line of the address of the Hospital [character]
5.	State	State in which hospital is located [character]
6.	Specialties	Specialties of the hospital [list of character]
7.	Facilities	Facilities available in the hospital [list of character]

3.2 Data Preprocessing

After the collection of data, it is preprocessed to improve the quality of the data. All the three datasets contained different sets of diseases so the common disease was selected to integrate them. The pre-preprocessing that is done on the three datasets is explained as follows:

1. **Symptoms Disease Dataset**
   There were no null values in the dataset shown in Table 4. There were 132 attributes in the dataset. We observed that only 91 attributes contributed to the prediction of disease and hence were selected for the training of the model.
2. **Drug Review Dataset**
   In the drug review dataset shown in Table 5, the review column was dropped and a new attribute which was a combination of rating and the useful count was added. A rank matrix was then formulated using this new attribute corresponding to the disease and drug. The drugs were defined in the rows and the diseases in the columns. The values in the rank matrix are normalized for each disease individually in the range 0–10.
3. **Hospital Directory Dataset**
   All the columns were dropped except Hospital_Name, Hospital_Category, Discipline_Systems_of_Medicine, Address_Origi nal_First_Line, State, Specialties and Facilities. The rows with null entries were dropped, shown in Table 6. Facilities attribute contain a list of facilities that a hospital has. This list was split and the number of facilities was counted and according to the number of facilities, a score was assigned to each hospital. The specialties attribute also contains a list for each hospital in which each specialty was matched with the list of diseases in the symptoms disease dataset. The data frame was transformed into one hot

Table 4 Sample of preprocessed symptoms disease data

Itching	skin_rash	nodal_skin eruption	Continuous _sneezing	Shivering	Prognosis
1	1	1	0	0	Fungal infection
1	1	0	0	0	Chicken pox
0	0	0	0	0	GERD

Table 5 Sample of preprocessed drug review data

	(vertigo) Paroymsal positional vertigo	Acne	AIDS
(vertigo) Paroymsal positional vertigo	0	0	0
Acne	0	0	6.009901
AIDS	0	0	0.083778

Table 6 Sample of preprocessed hospital directory data

Hospital _Name	Hospital _Category	Discipline_ Systems_of _Medicine	Address_ Original_ First_Line	State	T.B.	AIDS	Facilities _Count
Sai lee Hospital and Diagnostic Centre	Private	Allopathic	Prathamesh Horizon, New Mhb Colony, New Link Road, Borivali West	Maharashtra	0	1	5
Godrej Memorial Hospital	Public/ Government	Allopathic	Pirojshanagar, Vikhroli (East)	Maharashtra	0	1	21

encoded vector where 35 attributes, representing the diseases, were added. Each row, which represents a hospital, was 1 in the column for the disease in which the hospital specialized.

4 System Architecture

In this section, the overall architecture of the model and the development of a user interface is discussed. The proposed Smart Healthcare Model comprises a Rshiny web interface connected to a server. All the computation will be performed on the server side. On the server side, an R implemented machine learning model is executed. Results are generated and sent back to the web application through the server. Results include the disease predicted by the algorithm, drugs recommended for the predicted disease, and best hospital specialized in the predicted disease. The algorithms used in the prediction are decision tree, random forest, SVM, and artificial neural network. The user inputs the symptoms he is experiencing. The system will give the output of the disease predicted. Various methods are used to rank queries in the recommendation system. Top k query is one such approach which is based on efficient rank aggregation [20]. In this method, a different attribute of the dataset is aggregated or combined to generate a rate which can be used a basis for ranking in the system. Ultimately, top k rated products are provided to the users. Recommendation of drugs and hospitals is done by top k query method.

The system is divided into three modules:

1. Disease Diagnosis (on the basis of symptoms entered by the user)
2. Drug Recommendation (according to the diagnosed disease)
3. Hospital Recommendation (best-rated hospitals in the specific disease)

5 Design and Implementation of Proposed Model

The flow of the proposed system is shown in Fig. 1. The user enters the symptoms in the web application. The symptoms are sent to the server where computation is done. The server predicts the disease from the symptoms and sends it back to the user. The user can then see the drug recommendations and hospital recommendations for the disease predicted. The detailed description of each module, disease diagnosis, drug recommendation, and hospital recommendation, has been given in the following sections.

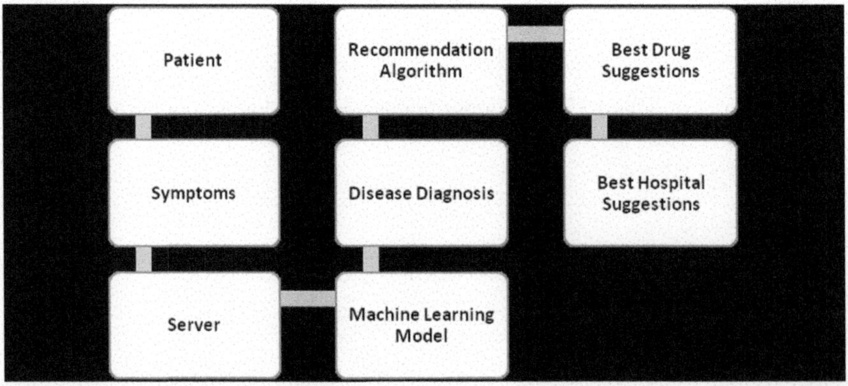

Fig. 1 Flow of the proposed model

5.1 Module 1: Disease Diagnosis

In this module, disease diagnosis is done on the basis of symptoms. The user is asked to select the five symptoms he is suffering from. On the basis of his selection, one hot encoded vector of symptoms is generated. This vector is passed to the server side which then feeds it to a pre-trained machine learning model. Decision Tree, Random Forest, SVM, and Artificial Neural Network algorithm have been used to train four different models. The prediction that is most common in all the four models is declared as the final prediction. The flow of the working of the module is shown in Fig. 2.

The algorithms used in the model are explained below.

- **Decision Tree**
  Decision tree is a supervised learning algorithm. It is used for both classification and regression problems. It identifies different ways to split a dataset based on certain conditions. It creates a tree-like structure that makes the prediction of a target variable. The nodes represent the split on some attribute. The reasons to use decision tree is that it works just like a human brain which is easy to understand and process. The tree helps us to visualize the actual working in the prediction of result. In a decision tree, the attribute is represented by nodes, the decision is represented by branches and the result is represented by leaf node [21]. Mainly two algorithms are used to construct a decision tree:

  1. CART (Classification And Regression Tree): Metric used is Gini Index
  2. ID3 (Iterative Dichotomiser 3): Metric used is Entropy function and Information Gain

  Gini index has been used as a metric for splitting on the features.
  Gini index measures the frequency of incorrect prediction of any randomly chosen element. In other words, it is used to measure how 'pure' the leaf nodes are. A

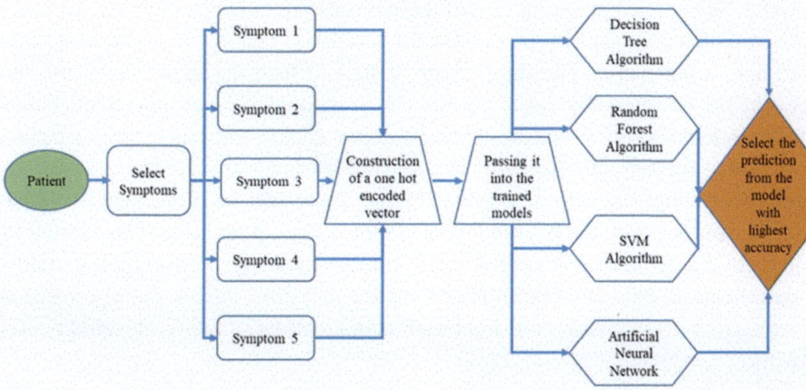

Fig. 2 Design of disease diagnosis algorithm

node having perfect class purity will have G = 0, whereas a node having G = 0.5 will have worst purity for binary classification. Hence, a node with least Gini index value would be preferred.

$$GiniIndex = 1 - \sum_{i=1}^{c} {p_i}^2 \tag{1}$$

Here c is the number of classes and pi is the probability of each class.

- **Random Forest**
  Random forest is an extension of the decision tree algorithm. It comprises multiple decision trees. Decision trees are created by randomly selecting some data samples from the dataset. Each decision tree is constructed using either CART or ID3 algorithm. For classification, each tree gives its own prediction and the final decision is made by majority vote. In regression problems, the average is taken of the outputs given by all the trees. More the number of trees the forest has, more robust it will be [7].
- **Support Vector Machine (SVM)**
  In an n-dimensional space each data item is plotted, where n is equal to the number of features. The value of each feature is depicted by the value of each coordinate. For classification, a hyperplane is found which will best segregate the two classes. The purpose of this algorithm is to find a plane which maximizes the distance between the data points of separate classes. The number of features helps to find the dimension of hyperplane. When the features exceed 3 the calculation of the dimension of hyperplane becomes tedious. Position of hyperplane is influenced by the support vectors, which are the points placed near to the plane. Deletion of support vector will lead to some modification of the hyperplane. These points are used in constructing a hyperplane. Hinge loss is the loss function which helps in maximizing the margin [22].
- **Artificial Neural Network (ANN)**
  Artificial Neural Network (ANN) is a model of interconnected neurons which compute values from some input. All the neurons are connected with each other using weighted connections. It contains three types of layers—input layer, hidden layer, and output layer. The input layer is used to feed the input value into the network. The number of nodes in the input layer is equal to the number of attributes or independent variables. The input layer does not change the data. It simply passes on the value further into the network. The hidden layer applies transformations to the input using the weighted connections. There can be more than 1 hidden layer. The output from the hidden layers is sent to the output layer. The number of neurons in the output layer is equal to the number of classes in the target variable. In classification, only one output node is active at a time. ANNs can also be used for regression problems. Figure 3 shows the architecture of a three-layered neural network.

Fig. 3 Neural network

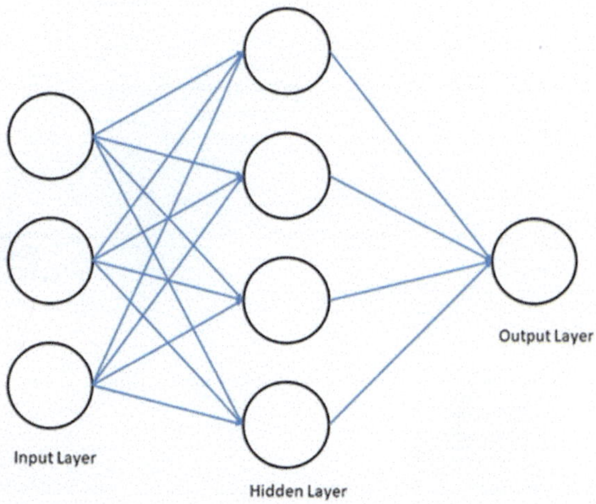

5.2 *Module 2: Drug Recommendation*

Raw and unfiltered data is loaded. Drugs are recommended on the basis of the overall rating. The overall rating of the drug is calculated using two attributes: rating and useful count. Useful count is the count of patients who have reviewed the drug as useful. Rating is the average of the rating given to the drug by the consumers. After the calculation of the overall rating, it is normalized into the range of 0–10. An 1147×34 dimension rank matrix is then constructed where the rows stand for the drugs and columns for the diseases. Top-rated drugs are suggested for the disease which was predicted in the first module. The flow of the drug recommendation is shown in Fig. 4.

5.3 *Module 3: Hospital Recommendation*

The third module (shown in Fig. 5) aims at suggesting the best-rated hospitals to the user. The raw data was first cleaned and preprocessed. The preprocessed data has 483 tuples and 42 attributes. Each row was converted to one hot encoded vector and contains a score representing the number of facilities provided by the hospital. The hospitals specializing in the disease predicted in the first module are sorted in decreasing order according to the score. The details of the hospitals having highest score are provided to the user. The user can filter out the hospitals according to his/her location.

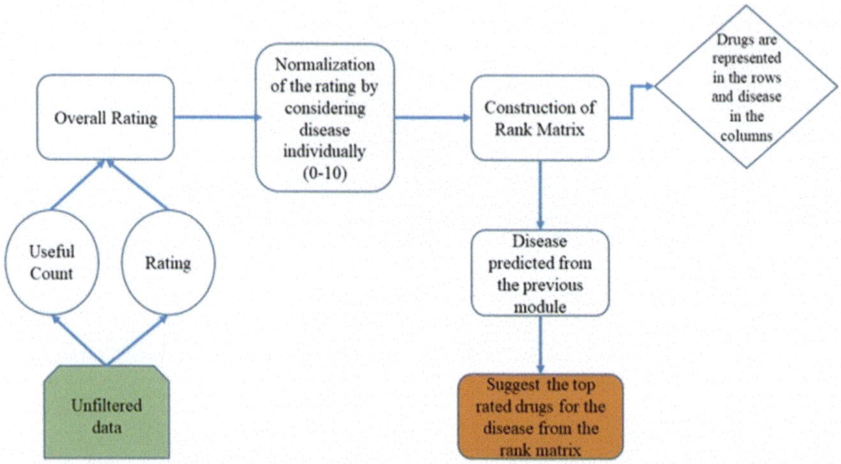

Fig. 4 Design of drug recommendation model

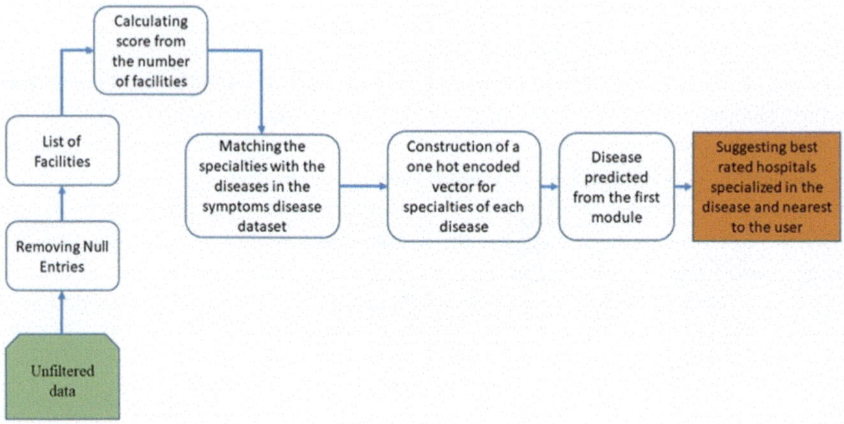

Fig. 5 Flow of hospital recommendation

6 Implementation Details

The research work includes two phases. First is on the development of backend for computation of results and the second is on the development of a user interface which will display the results. The backend part is implemented in R and a web application is made for the user interface. The web application is implemented using Rshiny. Rshiny is framework provided by Rstudio which is popular for its simplicity. It also

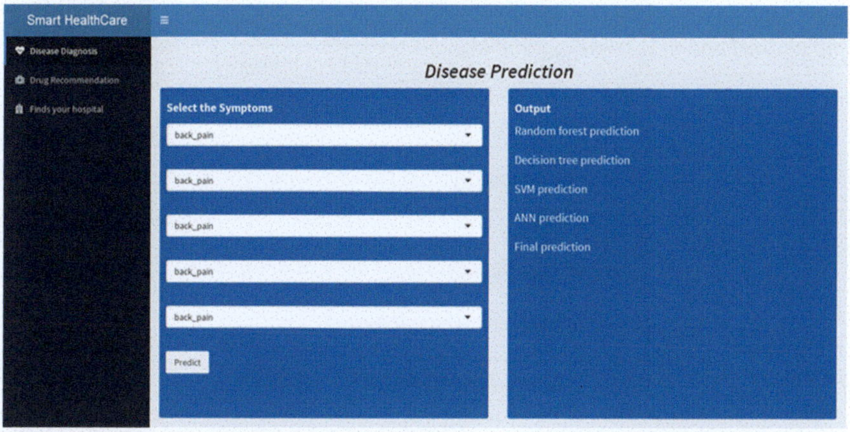

Fig. 6 Snapshot of disease diagnosis

allows live access to anyone. It caters to active binding of input and output. It provides wide-ranging widgets which makes the application interactive and impressive with basic efforts.

A shiny dashboard was built which shows the user three options—Disease Diagnosis, Drug recommendation, and Hospital recommendation. User can choose any tab according to his requirement. It provides an impressive and user-friendly framework to the developer. A dashboard page consists of dashboardheader, dashboardsidebar, and dashboardbody. It adjusts to the size of the screen automatically.

For disease diagnosis, four different machine learning models, decision tree, random forest, SVM, and neural net were trained and prediction from the classifier giving the highest accuracy was considered as the final output.

Figure 6 shows a snapshot of the disease diagnosis tab. Here the user can diagnose the disease that he might be having. The user is asked to select five symptoms that he is experiencing. When he clicks on the predict button, the diagnosis from each of the four models is shown.

The given pseudocode for drug recommendation elaborates on the implementation of the normalization of the ratings of the drugs and construction of the rank matrix. The final rating is brought down to the range of 1–10. It is done using the score and usefulcount. Then a rank matrix was constructed where drugs were on the rows and diseases on the columns.

$$normalizefunc \leftarrow \left(\left((max_{new} - min_{new}) * \frac{(x - min(x))}{max(x) - min(x)}\right) + min_{new}\right)$$

```
rating_normalized ← score * usefulcount
drugnorm ← data.frame( )
for( cond in diseaseunique)
        {  x ← drugbyclass[condition = cond]
    y ← apply(xrating, noramlizefunc, max_new = 10, min_new = 0
    y ← concatenating col x in y
        drugnorm ← concatenating row y in drugnorm
}
mat ← matrix(0, nrow = ndrug , ncol = ndisease)
rank_mat ← as.data.frame(mat)
colnames(rank_mat) ← disease_unique
rownames(rank_mat) ← drug_unique
for(drug in drug_unique){
    for(disease in disease_unique){
        a ← drugnorm[druname = drug and condition = dis]
        if(rating_normalized not empty){
            rank_mat[drug,dis] ← rating_normalized
        }
    }
}
prediction_drug{
  toprow ← selecting top 3 entries from sorted(descending) matrix
        }
```

Figure 7 is a snapshot of the dug recommendation tab. The user can get the best drug recommendations for the disease. When the user selects the disease and clicks on the “Suggest Drugs” button, the recommendations are displayed on the dashboard.

The pseudocode of hospital recommendation shows the calculation of the count of facilities, the attribute which was used for recommendation.

```
count_facilities ← data.frame(hospital, facility_count)
for(hosp in hospital)
        {
    fac ← selecting hospital facilities where name = hosp
    count_fac ← counting the number of facilities in fac
    a ← data.frame(hosp, count_fac)
    count_facilities ← concatinating row a in count_facilities
  }
```

The snapshot of the last tab is shown in Fig. 8. Here hospital suggestions are given to the user. The user enters the disease and by clicking on the “uggest Hospital!” button the top-rated hospitals for that disease is shown. For user’s convenience, along with the hospital name, category of the hospital, the system of medicine, address, and state is also given.

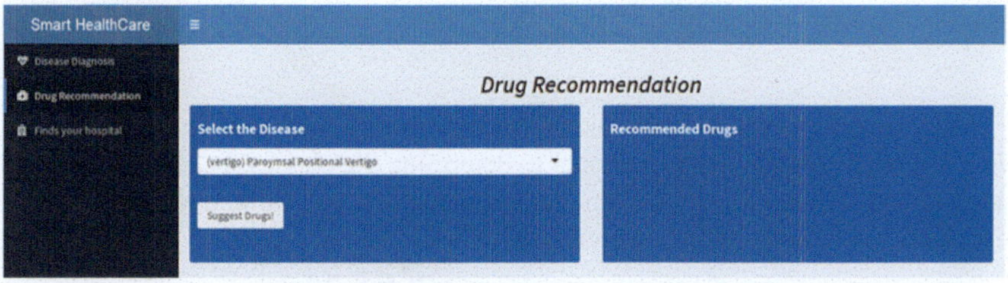

Fig. 7 Snapshot of drug recommendation

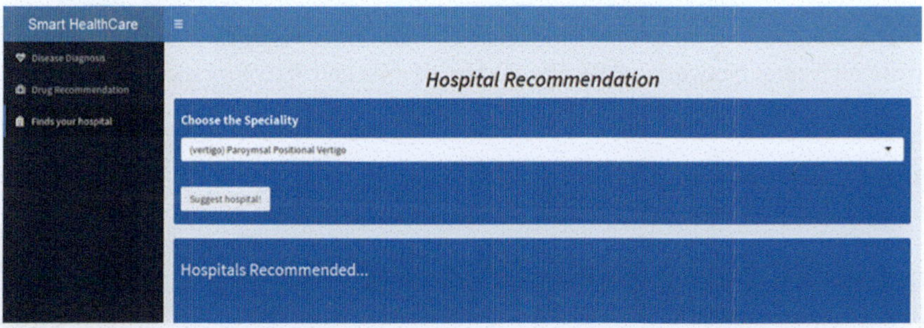

Fig. 8 Snapshot of hospital recommendation

7 Experimental Results

The dataset was divided into 75% training and 25% testing data. The training data was used to train the models. The models were then tested using the testing data. Precision, recall, F1 score, AUC value, and accuracy were calculated to compare the performance of the models.

$Precision$ is a measure of checking how often the model correctly predicts positive.

$$Precision = \frac{truepositives}{truepositives + falsepositives} \quad (2)$$

$Recall$ is the proportion of true positives which are correctly identified as positive. It also indicates that there are still some true positives which are identified incorrectly as negatives. A higher value of recall represents a higher value of true positive. For healthcare model like our higher sensitivity is desirable.

$$Recall = \frac{truepositives}{truepositives + falsenegatives} \quad (3)$$

$F1score$ is the harmonic mean of precision and recall.

$$F1score = 2 * \frac{precision * recall}{precision + recall} \tag{4}$$

AUC value is used to measure the performance of a model at different thresholds. It represents the capability of a model to determine the classes. Its value lies in the range of 0–1. Higher the AUC value higher is the distinguishing power. *Accuracy* represents how often the model is correct.

$$Accuracy = \frac{truepositives + truenegatives}{totalexamples} \tag{5}$$

The accuracy for all the four models Decision Tree, Random Forest, SVM, and Artificial Neural Network are 88.1%, 92.5%, 91.6%, and 98.6%, respectively. Table 7

Table 7 Results summary

Model name	Precision	Recall	F1 score	Area under ROC curve	Accuracy
Decision tree	0.951	0.891	0.884	0.978	0.881
Random forest	0.960	0.925	0.908	0.993	0.925
Support vector machine	0.961	0.919	0.904	0.990	0.916
Artificial neural network	0.987	0.985	0.985	0.996	0.986

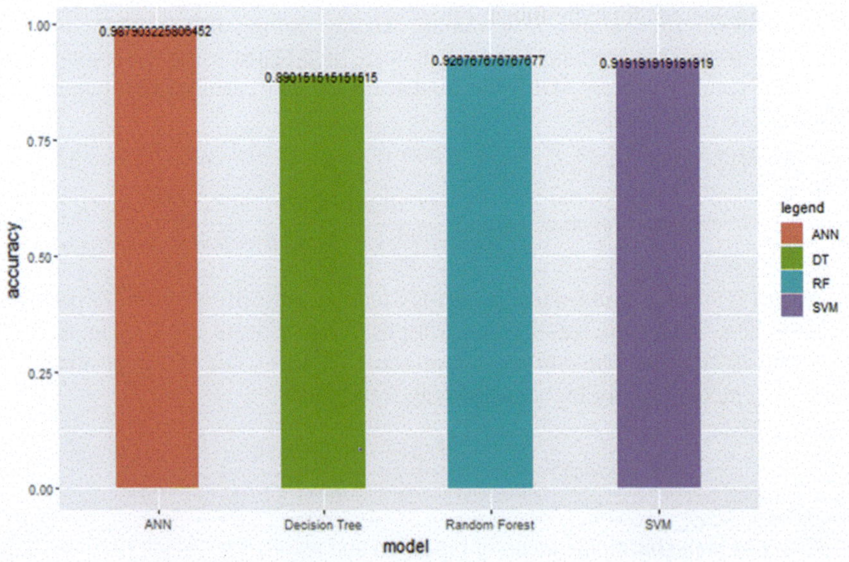

Fig. 9 Comparison of accuracy

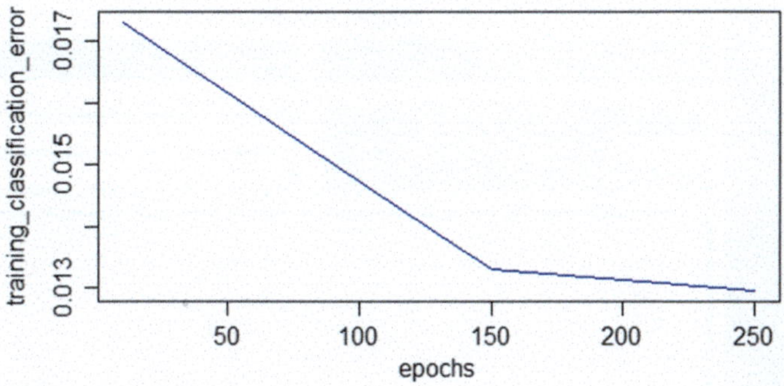

Fig. 10 Training scoring history of neural net

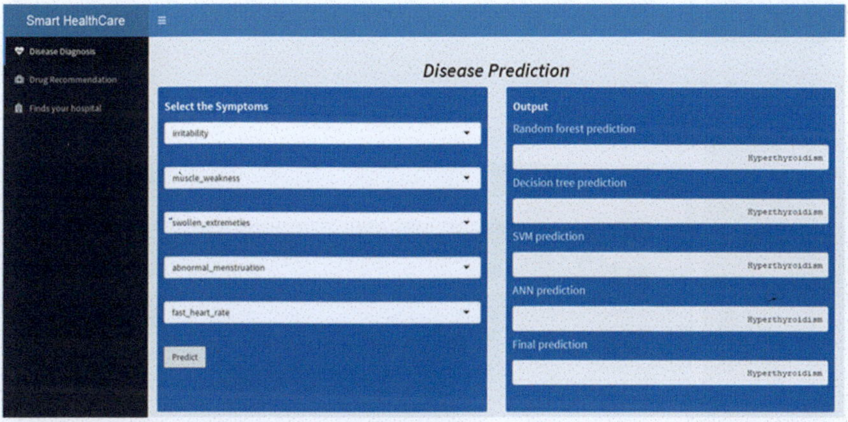

Fig. 11 Result of disease diagnosis

shows a summary of all the results calculated. From the results, it is clear that Artificial Neural Network gives the best results and hence is the best model. A graphical representation of the accuracies is shown in Fig. 9.

In Fig. 10, the training scoring history of the neural net is shown. As the number of epochs increase, the training error decreases. Till 150 epochs a steep negative slope can be seen. This means that the training error reduces at a high rate. 150 onwards the training error reduces gradually.

In Fig. 11, we can see that for irritability, muscle weakness, swollen extremities, abnormal menstruation, and fast heart rate the disease predicted is Hyperthyroidism.

For Hyperthyroidism, the best three drugs are Methimazole, Tapazole, and A+D Cracked Skin Relief. A graph of the unique number of drugs rated for each disease is shown in the graph. The snapshot of the result is shown in Fig. 12.

As shown in Fig. 13, the user can choose the hospital as per his location. For instance, the best hospital for Hyperthyroidism in Uttar Pradesh is Ojjus Medicare Goodwill Hospital and Research Centre.

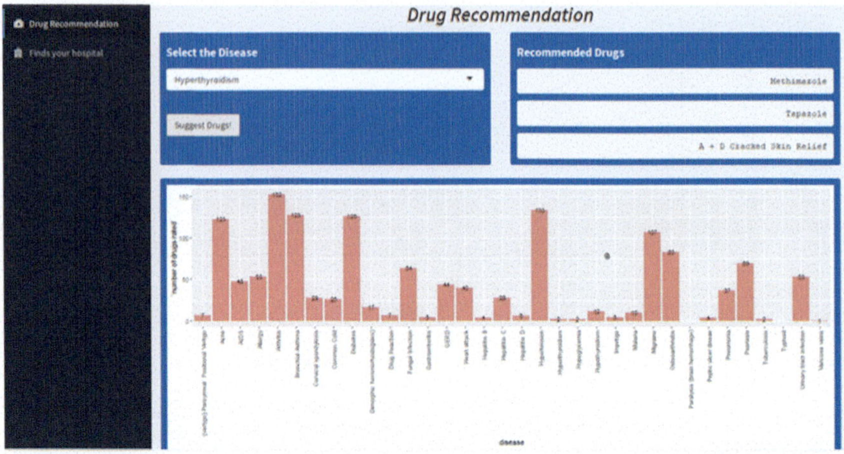

Fig. 12 Result of drug recommendation

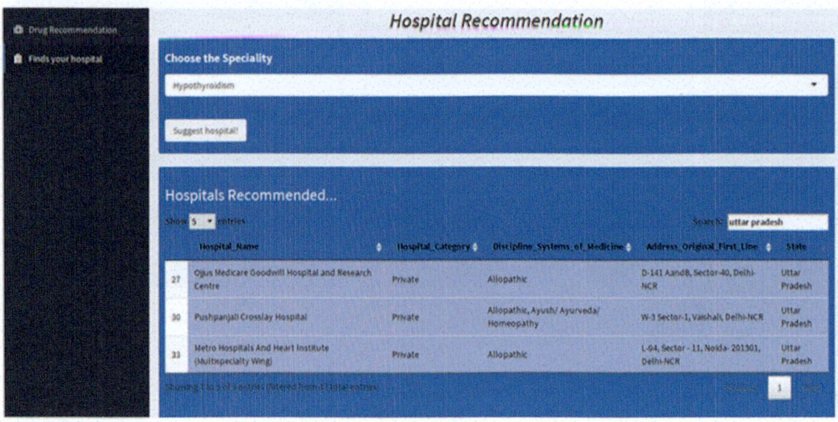

Fig. 13 Result of hospital recommendation

8 Conclusions

In this paper, we have tried to formulate an intelligent healthcare system which includes disease prediction, drug and hospital recommendation. Taking into consideration the relationship between the symptoms and various diseases, we have designed a system that would help in the prediction of diseases based on the symptoms entered by the user. This model was trained using different machine learning algorithms. Experimental analysis was done to evaluate each model. The final prediction that is given to the user is the one which is the most common in all the four models. The next part of the paper focuses on the recommendation of the drug corresponding to the disease predicted. It is done on the basis of rating and useful count of the drug provided by the users. A rank matrix containing the normalized rating between drugs and diseases was formed and then the recommendation of the drug was done. Ultimately a hospital was recommended to the user on the basis of specialties and facilities of the hospital. These recommendations are done by the top-k query. Experimental results indicate that the system is precise, efficient, and scalable.

9 Future Work

This system could be extended to consider the side effects that a specific drug might have on any patient. This can be done by taking insights from the patient's medical records. Relationship between two drugs can be established to prevent the side effects when they are taken together. Further hospital recommendation can also include an individual's insurance coverage. Recommendation of a healthy diet can also be done so that the chances of disease occurrence reduce.

References

1. N. Solhjoo, N. Naghshineh, F. Fahimnia, A.R. Ameri-naeini, Interventions to assist pet owners in online health information seeking behaviour: a qualitative content analysis literature review and proposed model. Health Inf. Libr. J. **35**(4), 265–84 (2018)
2. J.D. Birkmeyer, E.V. Finlayson, C.M. Birkmeyer, Volume standards for high-risk surgical procedures: potential benefits of the Leapfrog initiative. Surgery **130**, 415–22 (2001)
3. E. Gündoğan, B. Kaya A link prediction approach for drug recommendation in disease-drug bipartite network. In: 2017 International Artificial Intelligence and Data Processing Symposium (IDAP) (2017, IEEE), pp. 1–4
4. J.A. Knottnerus, P. Tugwell, Evidence-based medicine: achievements and prospects. J. Clin. Epidemiol. **84**, 1–2 (2017)
5. Y. Yesha, V.P. Janeja, N. Rishe, Y. Yesha Personalized decision support system to enhance evidence based medicine through big data analytics. In: 2014 IEEE International Conference on Healthcare Informatics, pp. 376–376 (2014)
6. C.Y. Chiang, C.T. Chang, R.E. Chang, C.T. Li, R.M. Huang, Patient and health system delays in the diagnosis and treatment of tuberculosis in Southern Taiwan. Int. J. Tuberc. Lung dis. **9**(9), 1006–12 (2005)
7. L. Breiman, Random forests. Mach. Learn. **45**(1), 5–32 (2001)
8. K. Li, C. Liu, K. Li, A.Y. Zomaya, A framework of price bidding configurations for resource usage in cloud computing. IEEE Trans. Parallel Distrib. Syst. **27**(8), 2168–81 (2015)
9. A. Cawsey, F. Grasso, C. Paris, Adaptive information for consumers of healthcare, *The Adaptive Web* (Springer, Berlin, 2007), pp. 465–485
10. R.C. Chen, Y.H. Huang, C.T. Bau, S.M. Chen, A recommendation system based on domain ontology and SWRL for anti-diabetic drugs selection. Expert Syst. Appl. **39**(4), 3995–4006 (2012)
11. J. Chen, S.S. Rathore, M.J. Radford, H.M. Krumholz, JCAHO accreditation and quality of care for acute myocardial infarction. Health Affairs **22**(2), 243–54 (2003)
12. J.D. Birkmeyer, A.E. Siewers, N.J. Marth, D.C. Goodman, Regionalization of high-risk surgery and implications for patient travel times. Jama **290**(20), 2703–8 (2003)

13. J.B. Dimick, S.R. Finlayson, J.D. Birkmeyer, Regional availability of high-volume hospitals for major surgery: many patients continue to undergo high-risk surgery at hospitals with inadequate experience in performing their procedure. Health Affairs **23**(Suppl2), VAR-45 (2004)
14. M.M. Ward, M. Jaana, D.S. Wakefield, R.L. Ohsfeldt, J.E. Schneider, T. Miller, Y. Lei, What would be the effect of referral to high-volume hospitals in a largely rural state? J. Rural Health **20**, 344–54 (2004)
15. F. Greaves, D. Ramirez-Cano, C. Millett, A. Darzi, L. Donaldson, Use of sentiment analysis for capturing patient experience from free-text comments posted online. J. Med. Internet Res. **15**, e239 (2013)
16. G. Cohen, M. Hilario, H. Sax, S. Hugonnet, A. Geissbuhler, Learning from imbalanced data in surveillance of nosocomial infection. Artif. Intell. Med. **37**, 7–18 (2006)
17. Symptoms Disease Data (2019), https://www.kaggle.com/neelima98/disease-prediction-using-machine-learning/
18. Drug Review Data (2018), https://archive.ics.uci.edu/ml/datasets/Drug+Review+Dataset+%28Drugs.com%29
19. Hospital Review Data (2016), https://data.gov.in/catalog/hospital-directory-national-health-portal
20. Y. Song, Z. Zhuang, H. Li, Q. Zhao, J. Li, W.C. Lee, C.L. Giles Real-time automatic tag recommendation. In: Proceedings of the 31st Annual International ACM SIGIR Conference on Research and Development in Information Retrieval (ACM, 2008), pp 515–522
21. A. Criminisi, J. Shotton, E. Konukoglu, Decision forests: a unified framework for classification, regression, density estimation, manifold learning and semi-supervised learning. Found. Trends Comput. Graph. Vis. **7**(2–3), 81–227 (2012)
22. J.A. Suykens, J. Vandewalle, Least squares support vector machine classifiers. Neural Process. Lett. 293–300 (1999)

The manufacturer's authorised representative in the EU is Springer
Nature Customer Service Centre GmbH, Europaplatz 3, 69115 Heidelberg,
Germany. If you have any concerns regarding our products, please
contact ProductSafety@springernature.com

Printed and bound by CPI Group (UK) Ltd, Croydon, CR0 4YY
27/07/2026
02177387-0001